Ian Cunha D'Amato Viana Dragaud
Luiz P. F. Assad
Luiz Landau

The resurgence of Cabo Frio (RJ) during the mid-Holocene

Ian Cunha D'Amato Viana Dragaud
Luiz P. F. Assad
Luiz Landau

The resurgence of Cabo Frio (RJ) during the mid-Holocene

Evaluation of the potential for occurrence using results from paleoclimate simulations

ScienciaScripts

Imprint
Any brand names and product names mentioned in this book are subject to trademark, brand or patent protection and are trademarks or registered trademarks of their respective holders. The use of brand names, product names, common names, trade names, product descriptions etc. even without a particular marking in this work is in no way to be construed to mean that such names may be regarded as unrestricted in respect of trademark and brand protection legislation and could thus be used by anyone.

Cover image: www.ingimage.com

This book is a translation from the original published under ISBN 978-3-330-77212-0.

Publisher:
Sciencia Scripts
is a trademark of
Dodo Books Indian Ocean Ltd. and OmniScriptum S.R.L publishing group

120 High Road, East Finchley, London, N2 9ED, United Kingdom
Str. Armeneasca 28/1, office 1, Chisinau MD-2012, Republic of Moldova, Europe
Managing Directors: Ieva Konstantinova, Victoria Ursu
info@omniscriptum.com

Printed at: see last page
ISBN: 978-620-8-64098-9

INDEX

THE RESURGENCE OF CABO FRIO (RJ) DURING THE MIDDLE HOLOCENE

Ian Cunha D'Amato Viana Dragaud, Luiz Paulo de Freitas Assad, Luiz Landau, Audalio Rebelo Torres Junior

SUMMARY

This study analyzed atmospheric and oceanic variables and the upwelling index in order to study the influence of atmospheric and oceanic dynamics on the upwelling process in the Cabo Frio (RJ) region during the mid-Holocene (MH) period. We used results from climate experiments for the MH and pre-industrial (PI) periods, defined by the *Paleoclimate Modeling Intercomparison Project Phase 2* (PMIP 2) using the coupled *Community Climate System Model Version* 3 (CCSM3) from the *National Center for Atmospheric Research* (NCAR). The results indicate that during the HM, the upwelling in the Cabo Frio region showed a seasonal behavior, being more intense in spring and summer and less intense in winter and autumn. Also noteworthy is the possibility of more intense upwelling processes occurring in Cabo Frio during the HM compared to the PI period, during the spring and summer seasons.

1 INTRODUCTION

During Earth's history, the planet's atmospheric composition and climate have evolved little by little as a result of external forcings such as solar radiation, orbital parameters and internal feedback mechanisms between the components of the climate system (Riebesell, 2004). Há 2.6 million years ago, cycles of climatic extremes began (glacial and interglacial episodes) which characterized the Quaternary period, covering the Pleistocene (between 1.806 million years and 11.5 thousand years ago) and Holocene (11.5 thousand years ago) epochs (Gradstein *et al.*, 2004). These glacial-interglacial cycles had a frequency of approximately 100,000 years during the last 500,000 years and were dominated by variations in the Earth's orbital parameters (Imbrie *et al.*, 1992). For a better understanding, see Table 1 for a schematic diagram of geological time.

Table 1: Schematic Chart of Geological Time (Source: http://www.unb.br/ig/glossario/. Accessed on 07/10/2012).

Eon	**It was**	**Period**	Season	Lower time limit
			Holocene	11.5± 0 ka
			Pleistocene	1.806± 0 ka
Phanerozoic	**Cenozoic**	**Neogene**	Pliocene	5.332± 0 ka
			Miocene	23.03± 0 Ma
		Paleogene	Oligocene	33.9± 0.1 Ma
			Eocene	55.8± 0.2 Ma
			Paleocene	65.5± 0.3 Ma
		Cretaceous	.	145.5± 4.0 Ma
	Mesozoic	**Jurassic**	.	199.6± 0.6 Ma
		Triassic	.	251.0± 0.4 Ma

	Paleozoic	**Permian**	.	299.0± 0.8 Ma
		Carboniferous	.	359.2± 2.5 Ma
		Devonian	.	416.0± 2.8 Ma
		Silurian	.	443.7± 1.5 Ma
		Ordovician	.	488.3± 1.7 Ma
		Cambrian	.	542.0± 1.0 Ma
Proterozoic	**Neoproterozoic**	.	.	1.0 Ga
	Mesoproterozoic	.	.	1.6 Ga
	Paleoproterozoic	.	.	2.5 Ga
Archaean	**Neoarchaean**	.	.	2.8 Ga
	Mesoarchean	.	.	3.2 Ga
	Paleoarchean	.	.	3.6 Ga
	Eoarchean	.	.	~3.85 Ga
Hadean	.	.	.	4.6(?)~ 3.85 Ga

This work investigates the period of the Middle Holocene (MH) (~6,000 years ago), when atmospheric levels of CO_2 were lower than those of the present (Raynaud *et al.* 1993) and the seasonal cycle of sunshine was modified due to differences in orbital parameters, leading to a natural climatic change and, therefore, of solar radiation incident on the planet's surface (Melo, 2007).

During the HM period, the solar radiation available at the top of the atmosphere was significantly different from the amount of radiation available today. During the Northern Hemisphere summer (around August), the Earth was closer to the sun (perihelion) than at present, changing the seasonal cycle of sunlight. Therefore, in the MW, the maximum amount of sunshine in the tropics occurred in August and the minimum in February. However, this change in sunshine is not exactly symmetrical with respect to the equator, due to the tilt of the Earth's rotational axis and eccentricity

(Liu, *et al.*, 2004). Changes in orbital parameters have been associated with a stronger seasonal cycle in the Northern Hemisphere, with an average of 5% more solar radiation in summer and 5% less in winter (Berger, 1978b), and a weaker seasonal cycle in the Southern Hemisphere, with an average of 5% less solar radiation in summer (Vettoretti, *et al.* 1998).

According to Hall and Valdes (1997), the HM is a relatively recent paleoclimate period, marked by natural climate change, for which the average climate characteristics are similar to those of the present climate. There are paleoclimate indicators for this period, which allow the results of numerical models to be validated. It also allows the results of this integration to be compared with standard simulations of the HM, carried out by the various world climate centers, using global climate models in the *Paleoclimate Modeling Intercomparison Project* (PMIP), representing a good test for evaluating the performance of numerical models of the general circulation of the atmosphere.

It is essential to describe the structure and variability of Sea Surface Temperature (SST) and surface winds on a seasonal and interannual scale (Taschetto and Wainer, 2002) in order to understand the impact of these parameters on the climate of the area where the upwelling occurs. According to Carbonel and Valentin (1999), upwelling is a mesoscale phenomenon forced by the action of winds along the continental margin, which, together with the Coriolis effect, induces Ekman transport from the water near the coast towards the ocean. This transport is compensated for by the rise of cold, nutrient-rich water from the lower layers to the euphotic layer, where photosynthesis takes place, resulting in high primary productivity and fish production. In terms of primary productivity, the Brazilian coast is an oligotrophic area (poor in nutrients) (Brandini, 2006). However, the Cabo Frio region is an exception due to the occurrence of the resurgence process (Carbonel, 2003; Mahiques *et al.*, 2005; Nagai *et al.*, 2009).

Upwelling zones and other regions of high sedimentation rate are considered important for understanding climatic variations during the Quaternary period (Mahiques *et al.*, 2005). These regions may indicate a series of processes related to environmental changes and sea level variations during this period. However, on the southwest Atlantic continental margin, little is known about the temporal fluctuation of primary productivity under different paleoclimatic conditions (Nagai *et al*., 2009; Mahiques *et al*., 2007).

Some studies have contributed to a better understanding of the atmospheric and oceanographic processes in the Cabo Frio upwelling system for paleoclimate periods, providing important information to help understand the variability of this complex system. However, studies with a regionalized approach have been based on paleoclimate indicators that allow inferences to be made about paleoclimate conditions, providing an "observational" view of the climate. As paleoclimate information is sparse and for specific periods, numerical modeling contributes to filling the information gap between the local and global scales, either through atmospheric models or models that couple ocean, atmosphere and/or other components. According to Melo and Marengo (2008), these models are useful for examining the relationship between the various forces and the respective responses of the climate system, and paleoclimate indicators can be used to validate the simulations.

The aim of this study is to use paleoclimate simulation results to investigate the influence of atmospheric and oceanic dynamics on the upwelling process in the Cabo Frio region (RJ) during the mid-Holocene period.

2. LITERATURE REVIEW

2.1. Characterization of the study area

A particular characteristic of the southeastern region of Brazil is the small area drained by rivers. These are short and carry little suspended material, most of which is deposited in estuaries and coastal lagoons (Martins and Coutinho, 1981). On the coast of Rio de Janeiro, the lagoon system is made up of an inner part dating back to the Pleistocene, with lagoons (e.g. Laguna de Araruama) connected to the Atlantic Ocean, and an outer part dating back to the Holocene, with smaller lagoons (e.g. Laguna Brejo do Espinho) isolated by sand barriers (Laslandes *et al.*, 2006). The Cabo Frio region has physiographic units made up of sandy coastal plains (restingas and dunes), stretches, lagoons and river deposits, low continental elevations, peninsulas and coastal islands (Scheel-Ybert, 2000).

With regard to the geometry of the coast and bathymetry, the southeast coast of Brazil shows important variations, with large variations being observed in the width of the platform. In the Cabo Frio region, the presence of a prominent cape is accompanied by an abrupt variation in the orientation of the coastline from northeast-southwest to east-west. In addition, there is an intense bathymetric gradient, where the 100m isobar is found 7km from the coast (Figure 2.1) (Rodrigues and Lorenzzetti, 2001).

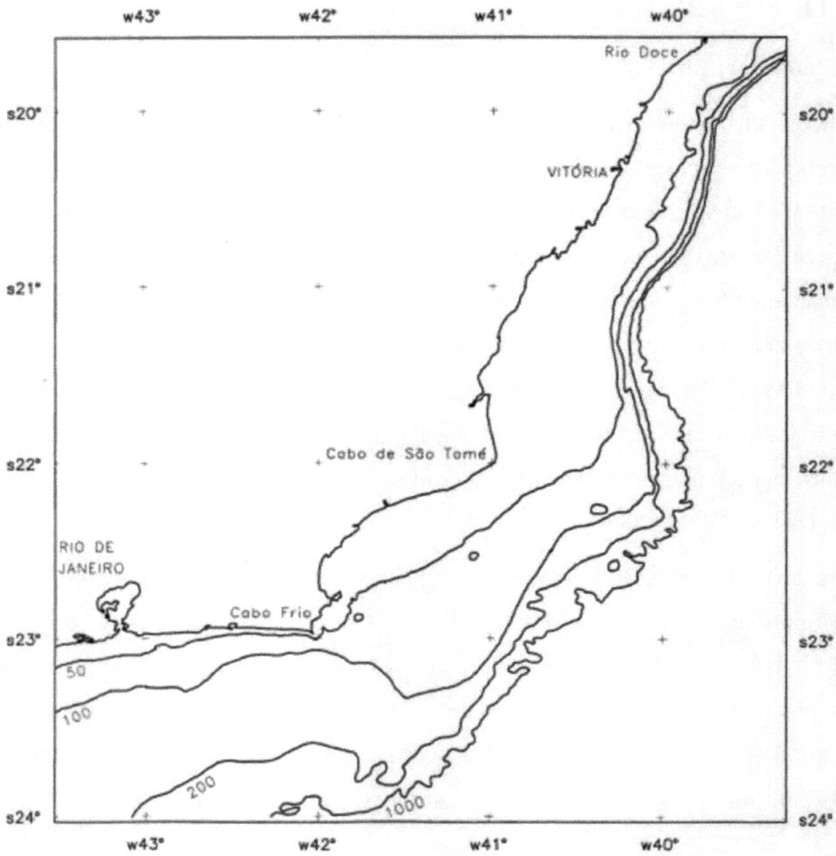

Figure 2.1: Southeast Brazilian continental shelf with isobars at 50, 100, 200 and 1000 m (Source: Rodrigues and Lorenzzetti, 2001).

As far as climate is concerned, most of the coastal zone of the state of Rio de Janeiro is typically humid, with high air temperatures and rainy in the summer, dominated by northeast winds, with a moderate dry season in the winter, when winds from the south quadrant linked to advections of colder air are more frequent. The Cabo Frio region has a particularly dry climate which is a variant of the hot semi-arid climate due to the resurgence process, which reduces local precipitation (Scheel-Ybert, 2000).

2.2. Cabo Frio's resurgence

According to Rodrigues and Lorenzzetti (2001), upwelling cells and plumes are often observed to the south of Cabo Frio, Cabo de Sao Tomé and near Vitória (Figure 2.2). However, the strongest signal (lower sea surface temperatures) occurs near Cabo Frio due to the upwelling, which is a marked phenomenon in this region and for this reason most studies of this phenomenon have been carried out in this area. Figure 2.2 shows the variation in TSM along the coast.

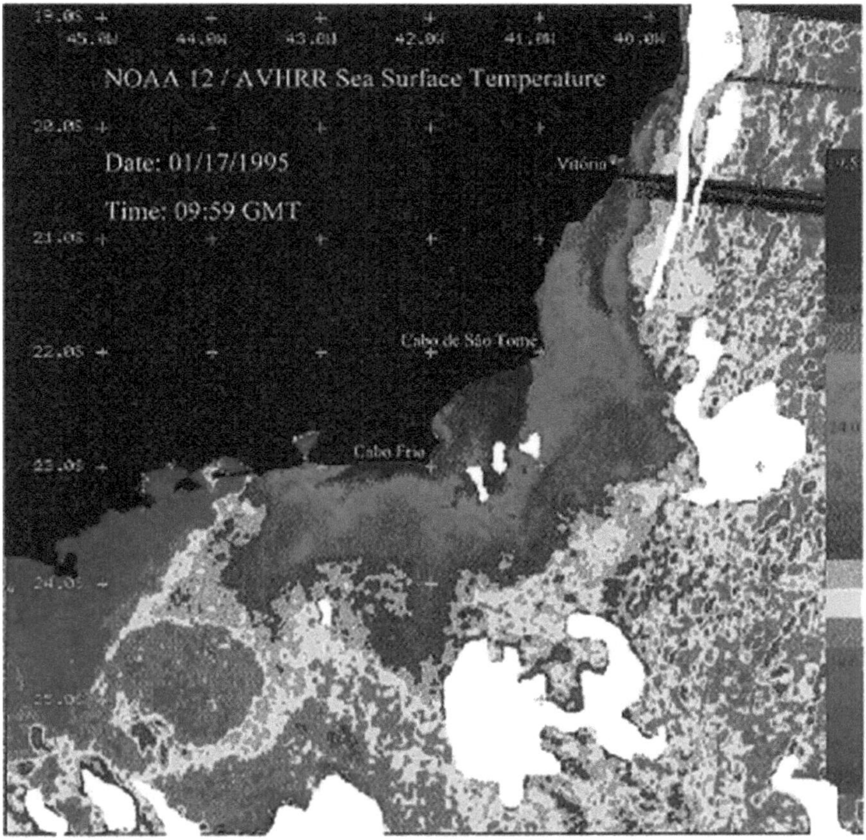

Figure 2.2: Satellite image of TSM (AVHRR-NOAA) on 17/01/1995 at 09:59 GMT, showing the upwelling on the continental shelf of Southeast Brazil. The clouds are marked in white and the values are in °C (Source: Rodrigues and Lorenzzetti, 2001).

According to Tanaka (1977), because the vertical movements of sea water in coastal areas are caused mainly by the influence of the winds, the intensity of the upwelling and its characteristics vary in each location and according to the time of year, depending mainly on the speed, direction and duration of the winds.

A body of *water* is defined as a body of water defined by characteristic values of temperature and salinity (T-S pairs), occupying a defined position in the water column and with a common history of formation. The upwelling in the Cabo Frio region is characterized as the rise of the mass of water known as the South Atlantic Central Water (SACW) resulting from the local action of atmospheric systems and therefore has a seasonal component, which implies a relationship with the South Atlantic Subtropical High (ASAS), being more intense in summer and weaker in winter (Torres Jr, 2005).

The ACAS is the source of waters with lower temperatures, which surface near the coast of Cabo Frio (Miranda, 1985; Valentin; André and Jacob 1987) and was characterized by Sverdrup (1942) with temperatures varying between 6°C and 18°C, and salinities between 34.5 and 36. More precisely, Miranda (1985) suggests 20°C and a salinity of 36.20 as the thermohaline index of the ACAS in the Southeast region of Brazil. This body of water is formed by the sinking of waters in the region of the Subtropical Convergence, and subsequent spreading along the surface of a density suitable for its hydrostatic adjustment (Silveira *et al.*, 2000). According to the compilation of works by Silveira *et al.* (2000), the ACAS enters as part of the Subtropical Gyre, circulates with the South Atlantic and Benguela currents, and reaches the coast of South America carried by the South Equatorial Current, flowing southwards below the latitude of 22°S.

The seasonal variation of the ASAS responds to variations in sunshine related to changes in the seasons and to continental temperature variations (Dominguez; Bittencourt and Martin 1992), which contributes to the seasonal variation of the upwelling in the Cabo Frio region, with the strongest signal being concentrated in the spring and summer seasons, due to the east-northeast winds coming from the ASAS during this part of the year. These winds have a component parallel to the coast, transporting surface water close to the continent towards the ocean and replacing it with water from greater depths (Valentin; André and Jacob 1987). During the autumn and winter seasons, the more frequent occurrence of cold fronts and a rapid succession of wind cycles shifting to the south quadrant are unfavorable to upwelling (Valentin; André and Jacob 1987). On the surface, the winds turn counterclockwise and flow for a few days in the southern quadrant (Franchito *et al.*, 1998).

The water column near Cabo Frio is usually characterized by a thermocline that fluctuates according to the intensity and direction of the winds (Valentin; André and Jacob 1987), and the intensity of the ACAS on the platform is also influenced by the morphology of the coast (Rodrigues and Lorenzzetti, 2001).

There is strong evidence of ecological and climatic impacts of the upwelling phenomenon in the Cabo Frio region (Maritin *et al.*, 1988; Franchito *et al.*, 1998; Laslandes *et al*, 2006), some of which are mentioned below.

Due to the occurrence of large-scale atmospheric subsidence in the subtropical gyres and the low humidity of the air during the upwelling seasons, the continental areas under the influence of this process suffer from the presence of stratus clouds and coastal fogs, and are characterized by desert climates and clear sky conditions (Tmartha, G. T., 1968, *apud* Bakun, 1990). The upwelling induces a reduction in local precipitation, an increase in evaporation and then in salinity, altering the local ecosystem (Martin *et al.*, 1988; Laslandes *et al.*, 2006). These conditions control the length of the dry seasons on the Fluminense coast, regulating the physical and chemical characteristics of the lagoon system and inducing the formation of hypersaline lagoons in the area influenced by the upwelling (eastern part of the coast) and hyposaline lagoons in the western part (Laslandes *et al.*, 2006).

Studies show that the characteristics of the upwelling process may be related to large-scale atmospheric processes, being controlled on a seasonal and interannual scale and by episodic events such as El Nino (Bakun, 1990; Martin *et al.*, 1988; Torres Jr., 2005).

Signs of the possibility of a connection between the Southern Oscillation (El Nino) and the winds that act locally in Cabo Frio, were raised based on studies of the sediments of the Araruama lagoon by Martin *et al.* (1988), proposing that prolonged phases of the Cabo Frio upwelling would be generated by strong El Nino events.

Torres Jr (2005) demonstrated that the teleconnection model proposed that the ASAS, in response to TSM anomalies in the equatorial Pacific via the Walker and Hadley circulation, influences the depth and consequently the availability of the ACAS for Ekman pumping in the southwestern South Atlantic basin. It was found that wind rotation in the western Atlantic region is related to El Nino events in the equatorial Pacific, with an increase in vorticity with a negative value in El Nino years, favoring Ekman pumping.

Bakun (1990) proposed a positive feedback between the effects caused by an increase in CO_2 levels and the upwelling process. Based on simulations of the increase in this gas, it was suggested that the low continental thermal pressures adjacent to the upwelling regions intensify, increasing the atmospheric pressure gradient and intensifying the wind along the coast, which in turn accelerates the upwelling circulation. However, the cooling of the ocean surface could result locally in an increase in the intensity of high pressures, which in turn would also increase the atmospheric pressure gradient, producing a positive feedback.

2.3. Paleoclimatic conditions associated with the Cabo Frio upwelling during the mid-Holocene in South America

During the Quaternary, the coastal evolution of east-northeast Brazil was controlled on a large scale mainly by sea level variations, winds and climatic variations (Figure 2.3) (Dominguez, Bittencourt and Martin 1992).

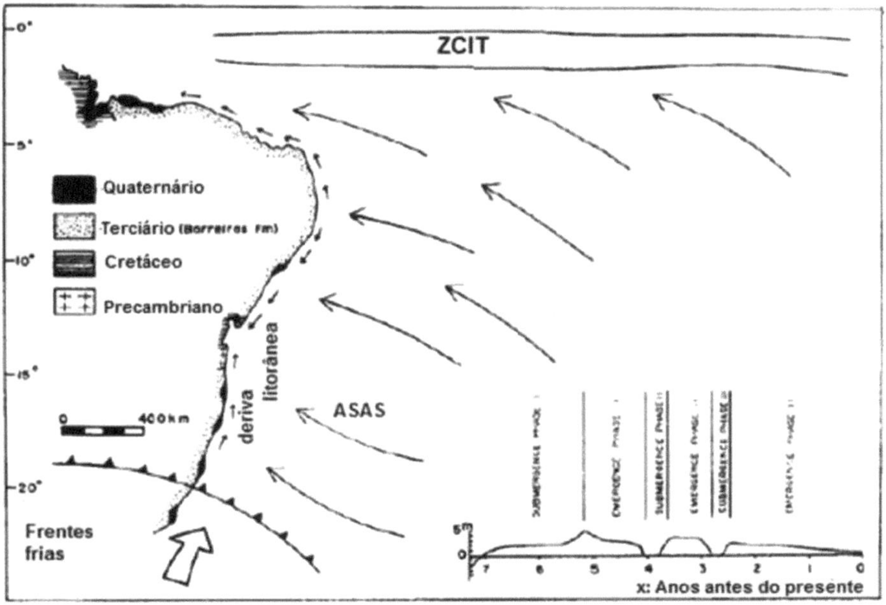

Figure 2.3: Simplified geological map of the east-northeast coastal zone of Brazil. Also shown are the major elements of the atmospheric circulation affecting the coastal zone and the sea level curve valid for the east coast. Sea level curve after Martin el. al, 1979 apud Dominguez, Bittencourt and Martin, 1992 (Source: adapted from Dominguez and Bittencourt and Martin, 1992).

Since the Pleistocene, the persistence of sediment dispersal in the east-northeast region of Brazil has been fundamentally controlled by the atmospheric circulation associated with the ASAS, which was remarkably stable during this period.

Despite this, this system has experienced small changes in position during the Holocene in response to climatic variations (Dominguez; Bittencourt and Martin, 1992).

Through simulations of the HM using the CCSM, Otto-Bliesner *et al.* (2006a) found that the main change in forcing for the HM is the variation in incident solar radiation. According to Figure 2.4, the radiation anomalies at high latitudes are large, 32 W/m^{-2} in the summer of the Northern Hemisphere and 48 W/m^{-2} in the spring of the Southern

Hemisphere. Annual average anomalies are small, less than 5 W/m^{-2}, suggesting modest annual impacts on the climate. Negative values occur in both hemispheres in December, January and February, with the lowest values occurring in the Southern Hemisphere. However, the area covered by the ocean surface is greater in the Southern Hemisphere than in the Northern Hemisphere, so the latter cools the most.

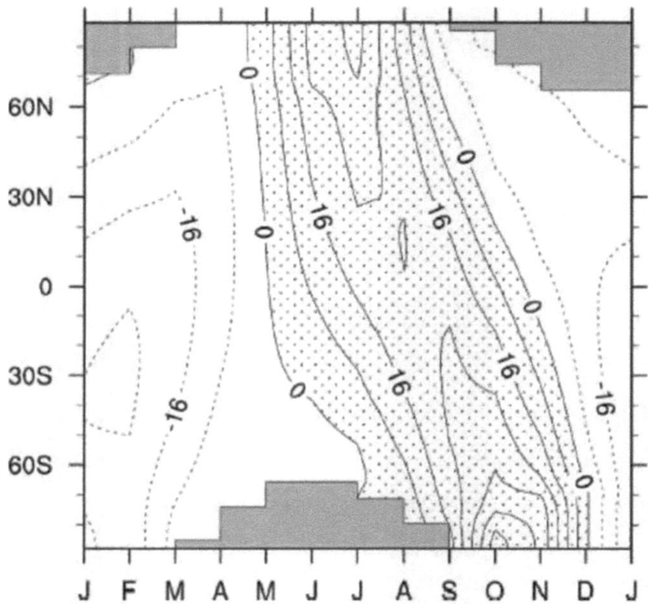

Figure 2.4: Temporal latitudinal distribution of the anomaly (mid-Holocene - IP) of solar radiation at the top of the atmosphere. The contour intervals are 8 W/m^{-2} (Source: adapted from Otto-Bliesner *et al.*, 2006a).

Melo and Marengo (2008) used the general atmospheric circulation model of the Centro de Previsao do Tempo e Estudos Climâticos (Center for Weather Forecasting and Climate Studies) to analyze climate variations during the HM, with a focus on South America. The amount of shortwave radiation (ROC) that reaches the top of the atmosphere during the HM shows that more radiation arrives during the austral and boreal summers and less during the winter months in both hemispheres, which is what characterizes the seasons. In the field of difference, it can be seen that during the MW the amount of radiation at the top of the atmosphere was different from today, due to the change in orbital parameters. The Earth was closer to the sun in the Southern Hemisphere winter during the HM, the exact opposite of today. As a result, more shortwave radiation (anomalously) arrived between April and September and less between October and March, between 0° and 30°S, de-intensifying the seasonal cycle of sunshine in the Southern Hemisphere tropics.

Melo (2007) identified a seasonal variation of the ASAS during the HM. In the summer, this system became more intense and closer to the continent. This pattern was similar to that observed in the fall and winter in the present climate. As a result, there was a de-intensification of the northeast trade winds, which may explain the northward shift of the Intertropical Convergence Zone (ITCZ). In the winter of the MW, there was a de-intensification of the ASAS, which could facilitate the spread of cold fronts, which were able to enter further north in South America and produce greater cooling than during the present day.

Figure 2.5 contains a summary of the main changes in the mean circulation at 850 hPa over South America during the summer, as identified by Melo (2007). It shows the intensification of the SSA circulation and the flow of the low-level jet from the SSA which is closer to the continent during the HM.

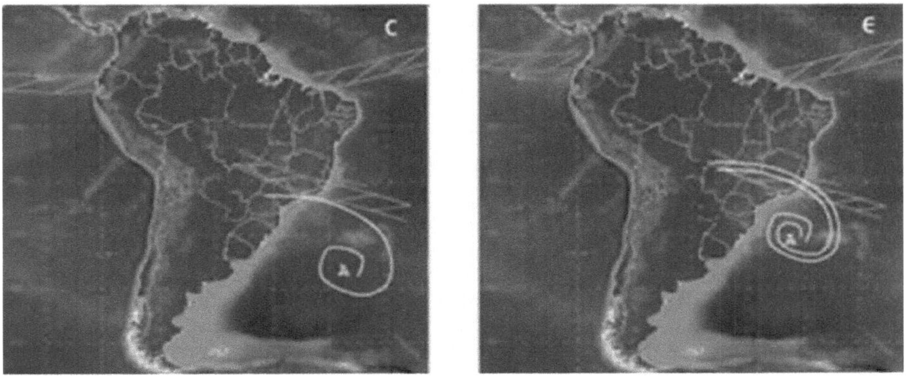

Figure 2.5: Summary of the main changes in the mean circulation at 850hPa over South America in summer. The figure on the left refers to the present climate and the figure

on the right refers to the mid-Holocene (Source: adapted from Melo, 2007).

Ledru *et al.* (1998) used pollen records and found that between 7000 and 4000 years ago humidity levels increased and apparently advection of polar air masses was less frequent.

During the Holocene, the sea level rose progressively until it reached its current average level 7,000 years ago. Subsequently, three main transgressive episodes occurred (7000-5100, 3900-3600, 2700-2500 years ago) interspersed with three regressive episodes (5100-3900, 3600-2700 and then 2500 years ago) (Martin *et al.*, 1993).

The intensity of El Nino during the Holocene underwent significant climatic variations (Laslandes *et al.*, 2006; Liu, Kutzbach and Wu, 2000). Laslandes *et al.* (2006), by analyzing a paleoclimate record from the Brejo do Espinho lagoon (eastern part of the coast influenced by the Cabo Frio upwelling) and comparing it with two other paleoclimate records, found that El Nino events were infrequent between 6500 and 4000 years ago, frequent between 3600 and 2900 and 2400 to 1800 and very frequent from this date onwards. These results are consistent with the results of HRM simulations by various authors, such as Melo (2007) who did not find El Nino and Otto-Bliesner *et al.* (2006a) who found a weaker El Nino, as well as Liu, Kutzbach and Wu (2000). These variations in the intensity of El Nino during the Holocene influenced the upwelling process in Cabo Frio, where prolonged phases of upwelling can be generated by strong El Nino events (Martin *et al.*, 1988).

Mahiques *et al.* (2007), in a study of the last 35,000 years off the coast of Brazil, investigated the influence of the BC in determining sedimentation in the region. It was suggested that during periods of higher relative sea level, the displacement of the BC towards the coast led to higher sea temperatures, which prevented an increase in productivity, as well as the deposition of organic matter (Figure 2.6). During periods when the sea level was lower, the movement of the BC away from the coast allowed for the discharge of material of terrestrial origin, as well as nutrients, favoring greater productivity and the deposition of organic carbon in the sediments (Figure 2.6).

However, exceptions to this pattern can occur in areas of upwelling, as described by Mahiques *et al*. (2005).

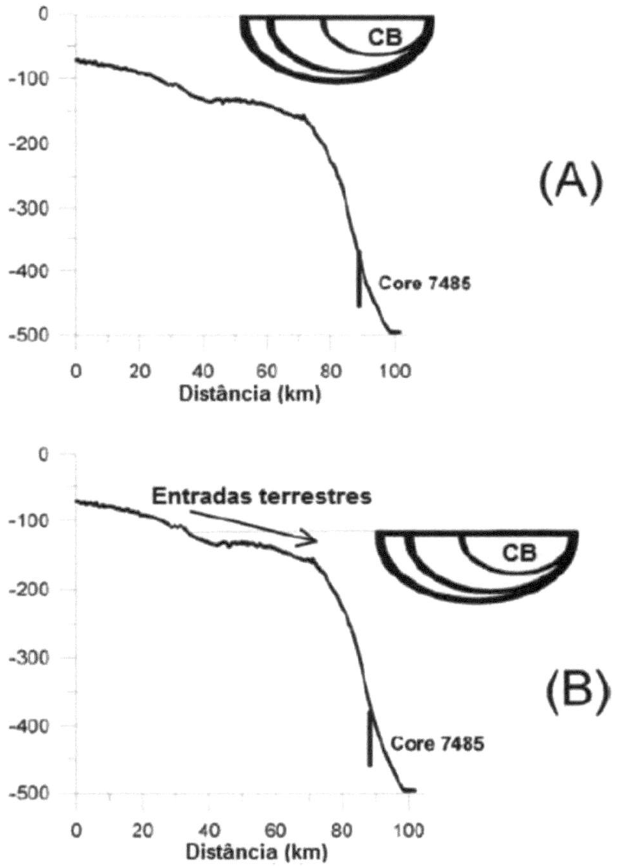

Figure 2.6: Schematic model for the position of the Brazil Current under different climatic conditions (a) higher sea level and (b) lower sea level (Source: adapted from Mahiques *et al.*, 2007).

Nagai *et al.* (2009), through the analysis of a seabed testimony, suggest that between 7,000 and about 5,000 years ago there was a strengthening of bottom currents, which according to Mahiques *et al*. (2007) this hydrodynamic condition is probably linked to the rise in sea level between 7,000 and 5,800 years ago. According to Nagai *et al.*

(2009), relatively high primary productivity as a result of the occurrence of ACAS on the shelf has also been suggested for this time interval. The presence of low organic carbon content was also identified between 7,000 and 5,000 years ago, and it is proposed by these authors that this must be related to one or more of the following: (1) resuspension of organic matter due to the relatively high velocity of the bottom current and the deposition of coarse sediments; (2) an increase in oxygenation, which would favor the degradation of organic matter; and/or (3) a decrease in the input of continental organic matter, due to high sea level conditions.

According to Laslandes *et al.* (2006), the intensity of drought in the lacunar context of the Fluminense coast is controlled by the occurrence of the Cabo Frio upwelling generated by northeast winds. The results found by these authors confirm that before 2400 years ago, frequent wet weather events were connected to predominantly south-southwest winds. After this date, northeasterly winds would be more intense, agreeing with the results of Manhiques (2005).

Manhiques (2005) investigated the current sedimentation in the Cabo Frio upwelling system and concluded that over the last 700 years a possible intensification of the ASAS and therefore of the north-easterly winds must have dynamically caused an intensification of BC transport and its mesoscale activity. Both strong and persistent northeast winds and CB meandering may have acted in combination to increase the upwelling process.

According to Nagai *et al.* (2009), from 2,500 years ago to the present, the upwelling process has intensified, controlled by the action of the northeast wind, the intensification of the meander pattern in the CB, and the morphology of the coast.

3. MATERIAL AND METHODS

The methodology of this work consists of analyzing atmospheric and oceanic variables and estimating the upwelling index (item 3.3) for the MH and PI climate experiments using monthly averages of the variables (item 3.2). The variables were analyzed using horizontal fields, zonal vertical sections in the ocean, temporal vertical sections in the ocean and point analysis of the upwelling index (item 3.3), with different coordinates for the dynamic and thermodynamic variables due to the different positioning on the model's numerical grid. For the thermodynamic variables, the sections were fixed at the 23.3°S parallel and the point analyses were carried out at longitude 41.7°W and the 23.3°S parallel. For the dynamic variables, the sections were fixed at the 23.5°S parallel and the point analyses were carried out at longitude 41.1°W and the 23.5°S parallel. The coordinates used for the point analysis consisted of the model's numerical grid, at the point closest to the Cabo Frio region.

To carry out this work, we used results from climate experiments for the MH and control period (PI), defined by PMIP 2 (item 3.1) using the NCAR coupled model, CCSM3 (item 3.4).

An overview of PMIP and information on the protocol used by PMIP 2 to simulate the coupled experiments for the MH are presented in section 3.1. More information about PMIP 2 can be found on the project's website (http://pmip2.lsce.ipsl.fr/. Accessed on: 07/10/2012).

3.1. THE PMIP

The *Paleoclimate Modeling Intercomparison Project Phase 1* (PMIP 1) was initiated in order to coordinate and encourage the systematic study of general circulation models of the atmosphere and to evaluate their ability to simulate major changes in climate such as those that occurred in the distant past. The project is centered on the periods of the mid-Holocene (6,000 years ago) and the last glacial maximum (21,000 years ago), because the climatic conditions were remarkably different in these periods, relatively large amounts of paleoclimate data existed and the main forcings are relatively well known (Jossaume, S; Taylor, K. E. 1995).

PMIP 2 addresses the role of different feedbacks using coupled models (Crucifix, M; Braconnot, S.P. and Otto-Bliesner, B., 2005), being in better agreement with paleoclimate data compared to PMIP 1, which was conducted with fixed TSMs under present-day conditions (Braconnot *et al.*, 2007). To avoid problems related to the prescription of TSM on a global scale, a major effort has been made by several groups in recent years to produce coupled simulations in general atmospheric-ocean circulation models (Ramstein, 2007), such as the experiments recently carried out by PMIP 2 (Braconnot *et al.*, 2007).

In order to determine the response of the ocean-atmosphere system to solar forcing, simulations were carried out with versions of models in which the vegetation is the same as in the control simulation (ocean-atmosphere coupled models) (Braconnot *et al.*, 2007).

3.1.1. The PMIP experimental protocol

As described by Braconnot *et al. (2007)*, the PMIP simulations follow a rigorous protocol, which represents the best compromise between the need to explain the different forces and realistic boundary conditions, and ensuring the relevance of the comparison between models and data.

Table 2. Boundary conditions and trace gases as recommended by the PMIP 2 project (adapted from Braconnot *et al.*, 2007).

	Layer of ice	Coastal Topography	Initial state of the ocean	CO2 (ppmv)	CH4 (ppbv)	NO2 (ppbv)
PI	Modern	Modern	Modern	280	760	270
MH	Same as 0k	Same as 0k	Same as 0k	280	650	270

Table 3. Terrestrial orbital parameters as recommended by the PMIP 2 project (adapted from Braconnot *et al.*, 2007).

	Eccentricity	Obliquity (°)	Angular

			Precession (°)
PI	0.0167724	23.446	102.04
MH	0.018682	24.105	0.87

3.1.2. Pre-Industrial Simulation

The reference (control) simulation is the pre-industrial (PI) climate pattern (around the 1750s). The orbital parameters are set to reference values from the 1950s, and the trace gases correspond to values from the 1750s (Tables 2 and 3). The difference between 1750 and 1950 sunshine induced by variations in orbital parameters is negligible, which explains why this aspect should be neglected. Small differences in solar forcing arise, however, from variations in the solar constant (Braconnot *et al.*, 2007). In the simulations presented by Braconnot *et al.* (2007), these variations are smaller than the variations caused by changes in insolation or boundary conditions. PMIP 2 uses the same solar constant for all experiments. Temperature and salinity are initialized from the Levitus (1998) dataset (http://pmip2.lsce.ipsl.fr/. Accessed on: 07/10/2012).

3.1.3. Mid-Holocene simulation

MH simulations make it possible to examine the climatic response to a variation in the seasonal and latitudinal distribution of incident solar radiation (insolation) caused by variations in orbital forcing (Berger, 1978a).

In the MH experiment, the most important change in the boundary conditions compared to PI is the variation in the pattern of insolation, due mainly to the precession of the equinoxes (Table 3). For simplicity and to help isolate the effects of the variation in the pattern of sunshine, it is assumed that the TSM prescribed for the HM is identical to the TSM of the present (Jossaume, S; Taylor, K. E., 1995).

Surface conditions and SSTs are considered to be the same as present-day values, due to the lack of data for this period, and also because HM SSTs are similar to present-day conditions, i.e. a pre-industrial atmosphere (Valdes, 2000).

The PMIP 2 protocol also considers a reduction in the concentration of CH4 (methane), with a lower concentration for MH. The concentrations of the other trace gases (CO_2

and NO_2) are maintained as in the control simulations (Table 2). Dust and other aerosols (volcanism) are not taken into account (Braconnot *et al.*, 2007).

3.2. Averages and Anomalies

Seasonal variations between the HM and IP periods are analyzed for three-dimensional atmospheric and oceanic variables. To do this, quarterly averages were calculated to represent each season of the year, using results from 10 representative years of the HM and IP. December, January and February represent the summer season. March, April and May represent the fall season. June, July and August represent the winter season. September, October and November represent the spring season. The 10-year time interval analyzed is due to restricted access to only results from this 10-year interval. In order to compare the two periods analyzed, anomalies were calculated between the seasonal averages of the variables of interest to the study, between the MH and PI periods. Thus, the anomaly is equal to the seasonal average of a variable during the HM minus the seasonal average of this variable during the IP.

Anomaly= average variable during MH - average variable during PI

3.3. The Resurgence Index

Ekman pumping results from the divergence of transport in the surface Ekman layer in the ocean caused by the spatial variation of the wind shear stress field (*Gill*, 1982). Applying the principles of conservation of mass and momentum, the vertical velocity at the base of the Ekman layer in the ocean (w) resulting from the frictional forces of the wind can be obtained by vertical integration of the continuity equation in the Ekman layer. The following formulation is applied as an index of upwelling.

$$w = \boldsymbol{k}.(\boldsymbol{\nabla} \times \frac{\tau}{f\rho}),$$

Where w is the vertical velocity at the base of the Ekman layer in the ocean, $\boldsymbol{k}$ is the vertical unit vector, $\boldsymbol{\tau}$ is the wind shear stress, f is the Coriolis parameter and ρ is the density of sea water.

In other words, the vertical velocities in the ocean are proportional to the rotational shear stress of the wind over the sea and inversely proportional to the product of the local planetary vorticity (Coriolis parameter) and the density of the sea water.

3.4. The NCAR climate model CCSM3

The *National Center for Atmospheric Research* (NCAR) *Community Climate System Model Version* 3 (CCSM3) is a global climate model that combines the atmosphere, ocean, sea ice and land surface (Collins *et al.* 2006b). The atmospheric component of this model is the third version of the NCAR *Community Atomspheric Model* (CAM3). CAM3 is a three-dimensional model, with primitive equations, solved horizontally using the spectral method, and has 26 levels vertically with a hybrid coordinate (Collins *et al.* 2006a). For these paleoclimate simulations, the atmospheric resolution is T42 (equivalent to a grid spacing of approximately 2.8° in latitude and longitude) (Otto-Bliesner *et al.*, 2006a). The land model uses the same grid as the atmospheric model and includes multiple land covers and plant functional types (Dickinson *et al.* 2006). The ocean model is NCAR's implementation of the *Parallel Ocean Program* (POP). POP is a three-dimensional model of primitive equations solved vertically with the *z* coordinate (Gent *et al.* 2006). For this simulation, the ocean grid has 320x384 points with the poles located in Greenland and Antarctica and 40 levels extending to a depth of 5.5km. The horizontal resolution corresponds to a grid spacing of approximately 1° in latitude and longitude with better resolution in the tropics and North Atlantic (Otto-Bliesner *et al.*, 2006a). The sea ice model is a dynamic-thermodynamic formulation (Briegleb *et al.* 2004), and uses the same horizontal grid and land mask as the ocean model (Otto-Bliesner *et al.*, 2006a).

The sensitivity of CCSM3 was studied by Otto-Bliesner *et al.* (2006a) through simulations of the Last Glacial Maximum and the HM. The results were compared with PI simulations. The HM simulation indicated an annual global cooling of less than 0.1°C compared to the PI simulation. Much larger and significant variations occurred regionally and seasonally, including a more intense summer monsoon in northern Africa, a reduction in Arctic sea ice in all months and weaker El Nino variability.

4. RESULTS AND DISCUSSION

The aim of this study was to use paleoclimate simulation results to investigate the influence of atmospheric and oceanic dynamics on the upwelling process in the Cabo Frio region (RJ) during the mid-Holocene (MH) period (~6000 years ago). To this end, oceanic and atmospheric variables were analyzed, an index of upwelling in the Southwest Atlantic region was calculated, and anomalies between the MH and pre-industrial (PI) periods were calculated in order to identify differences between these periods.

In the MW, atmospheric CO_2 levels were lower than at present (Raynaud *et al.* 1993) and the seasonal cycle of insolation was modified due to differences in orbital parameters, leading to natural climate change and therefore changes in solar radiation incident on the planet's surface (Melo, 2007). The seasons were characterized, with more ROC falling during the summer and less during the winter for both hemispheres. However, more ROC (anomalously) arrived between April and September and less between October and March, between 0° and 30°S, de-intensifying the seasonal cycle of insolation in the Southern Hemisphere tropics (Melo, 2007).

4.1. Sea water temperature

4.1.1. Surface fields

Figures 4.1a), 4.2a), 4.3a) and 4.4a) show the average TSM fields for the PI period for summer, fall, winter and spring respectively. It can be seen that the meridional temperature gradients become greater with increasing latitude, due to the decrease in solar radiation towards the South Pole. Higher temperatures can be seen near the coast due to the transport of water from the Brazil Current. However, during the summer and fall, a TSM gradient and lower values near the coast can be seen, as a result of the influence of the upwelling process. Figures 4.1b), 4.2b), 4.3b) and 4.4b) illustrate the anomaly of the SST field for summer, fall, winter and spring respectively. Negative values or values close to zero can be observed in the summer, fall and winter periods over the entire area studied. However, for spring, only positive values or values close to zero can be seen. Figure 5.10 illustrates the vertical temporal distribution of the sea temperature anomaly, in which we can see negative values on the surface from January to September. These negative sea temperature anomaly values indicate that lower temperatures occurred during the MH period compared to the PI period.

a)

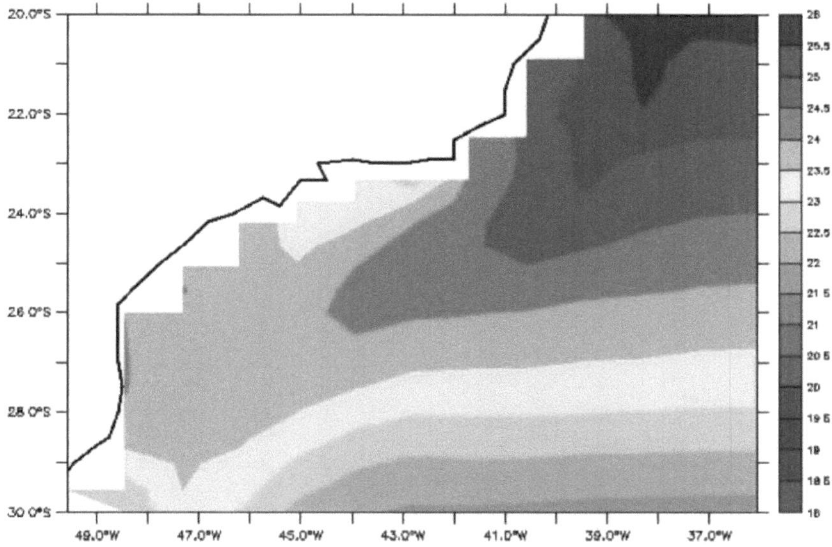

b)

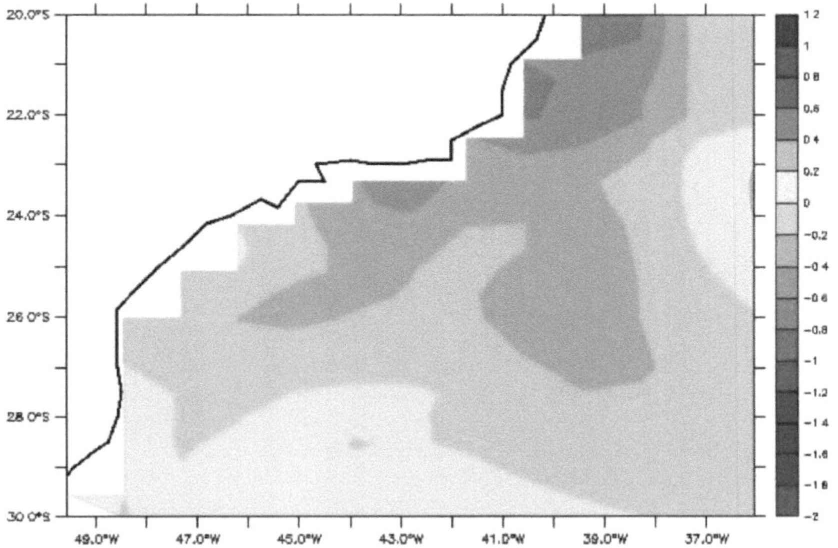

Figure 4.1: TSM field (°C) for: a) summer PI period, b) anomaly (MH-PI).

a)

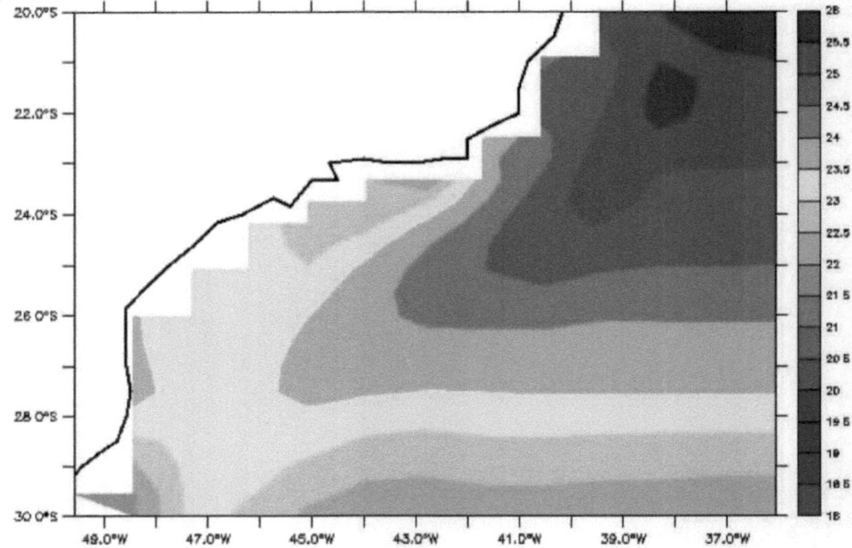

b)

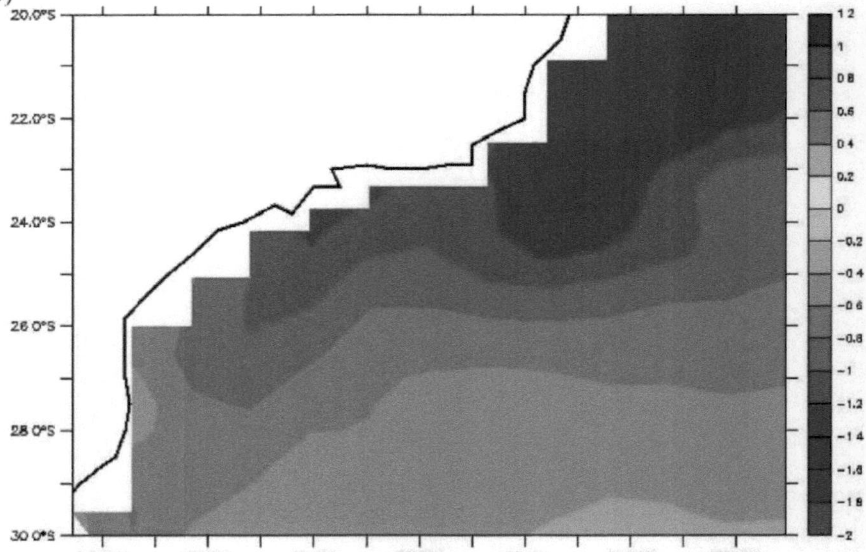

Figure 4.2: TSM field (°C) for: a) fall PI period, b) anomaly (MH-PI).

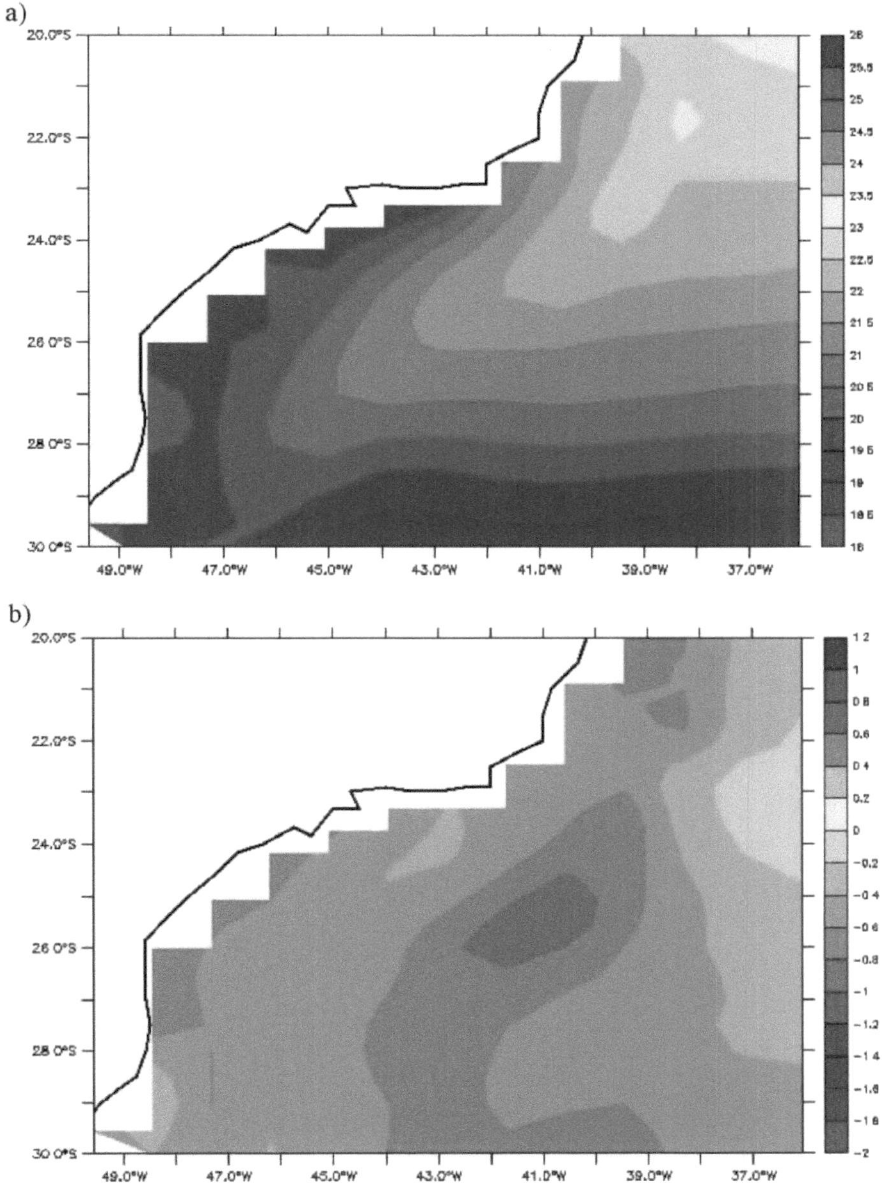

Figure 4.3: TSM field (°C) for: a) winter PI period, b) anomaly (MH- PI).

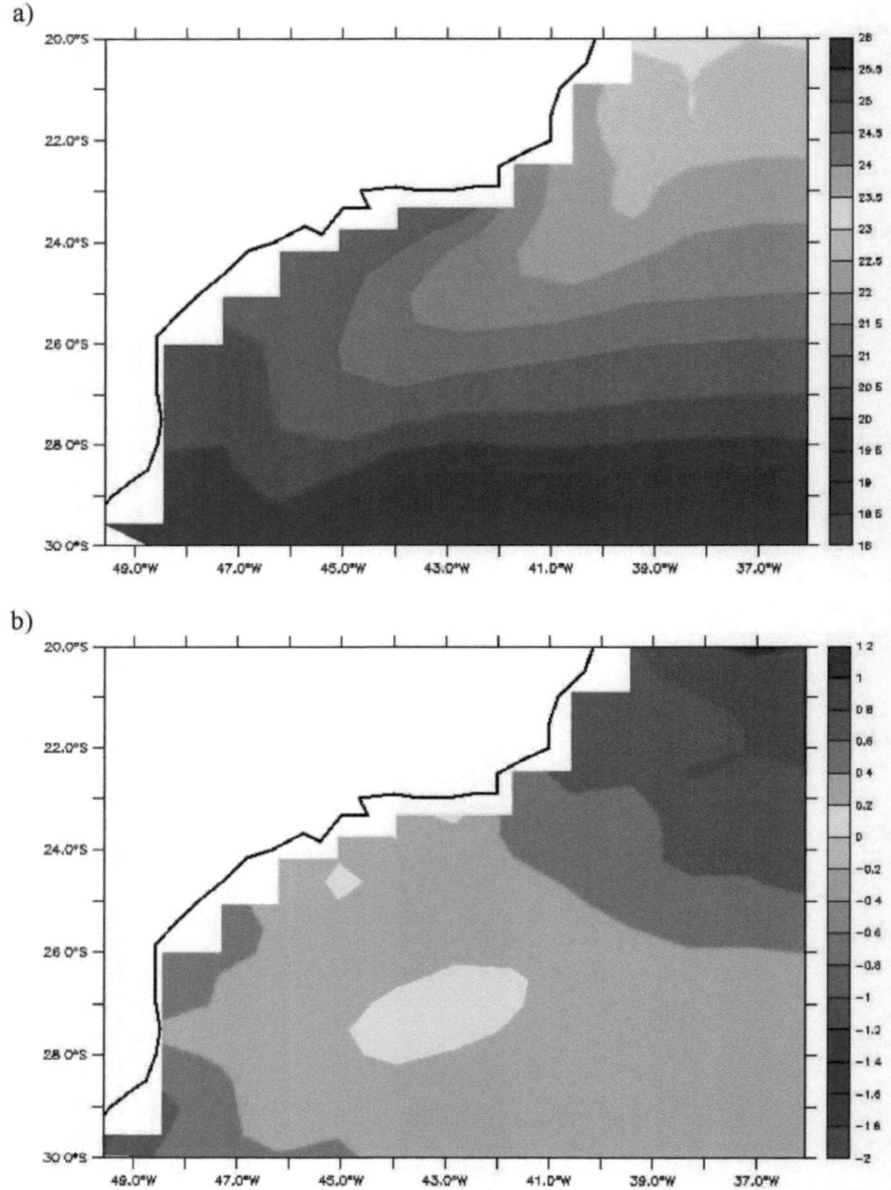

Figure 4.4: TSM field (°C) for: a) spring PI period, b) anomaly (MH- PI).

4.1.1 Vertical time distributions

The vertical time distributions of sea temperature in the ocean (Figure 4.5 a) and b)) show that the highest values occur between January and April and the lowest values between July and September. However, during the MH the maximum values are present over a shorter period and the minimum values over a longer period compared to the PI period.

a)

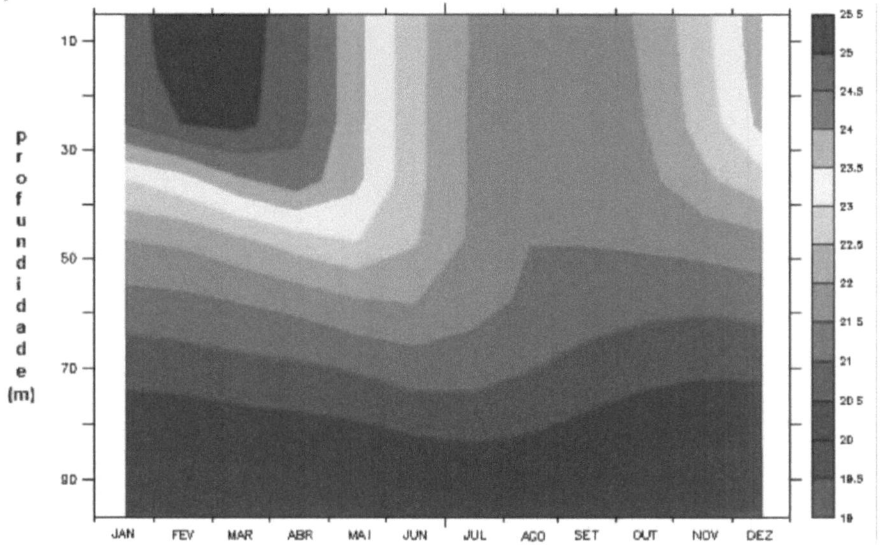

b)

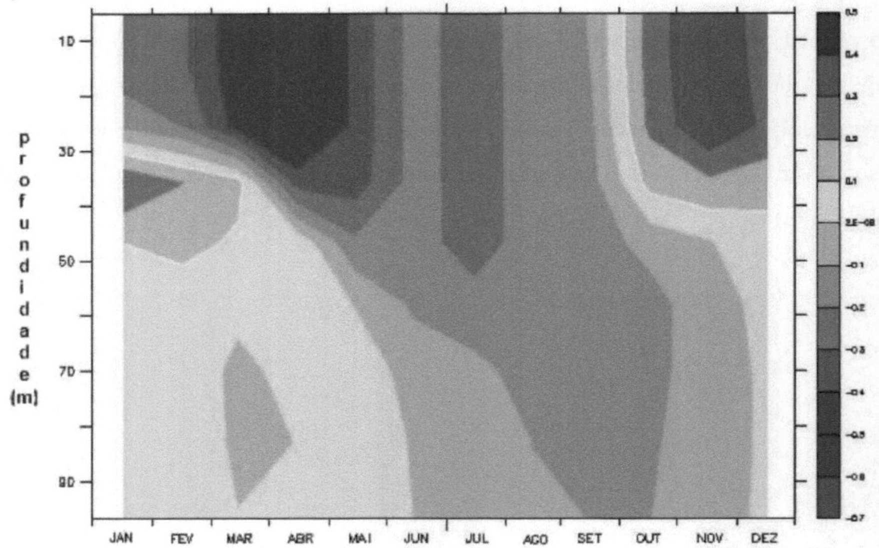

Figure 4.5: Vertical time distribution in the ocean of sea temperature (C°) for: a) PI period, b) anomaly (MH-PI).

4.2. Vertical component of the current

4.2.1. Zonal vertical sections

Figures 4.6 to 4.9 show the zonal vertical distributions of the vertical velocity of sea water for the 230 S parallel. More intense positive values can be seen in regions closer to the coast, which means a greater vertical upward velocity in this region. This may be related to the greater divergence of horizontal transport in this region due to the very presence of the coast and also to the greater occurrence of the upwelling phenomenon in regions close to Cabo Frio (RJ).

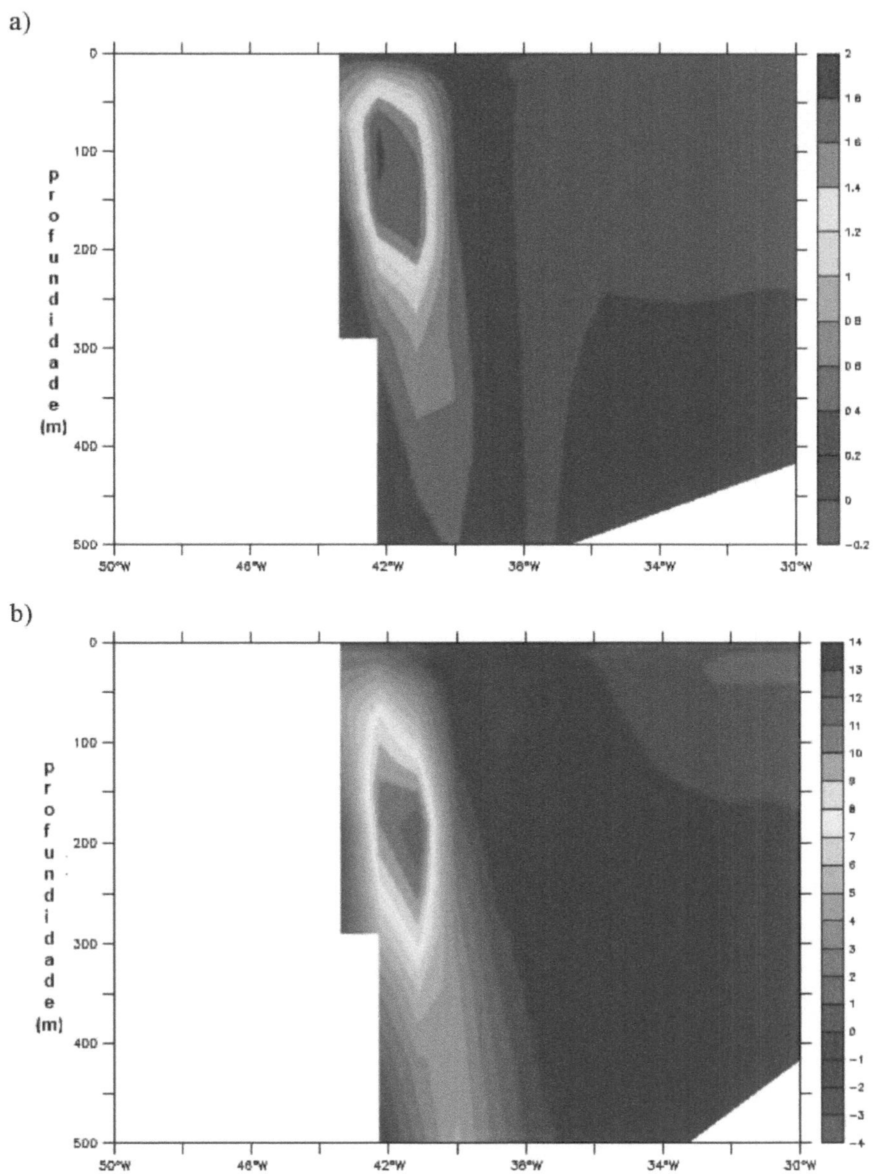

Figure 4.6: Zonal vertical section in the ocean of the vertical velocity of sea water ($m.s^{-1}$) for: a) summer of the PI period, b) anomaly (MH-PI). The scales correspond to values that have been multiplied by $10e^5$ and by 10, respectively.

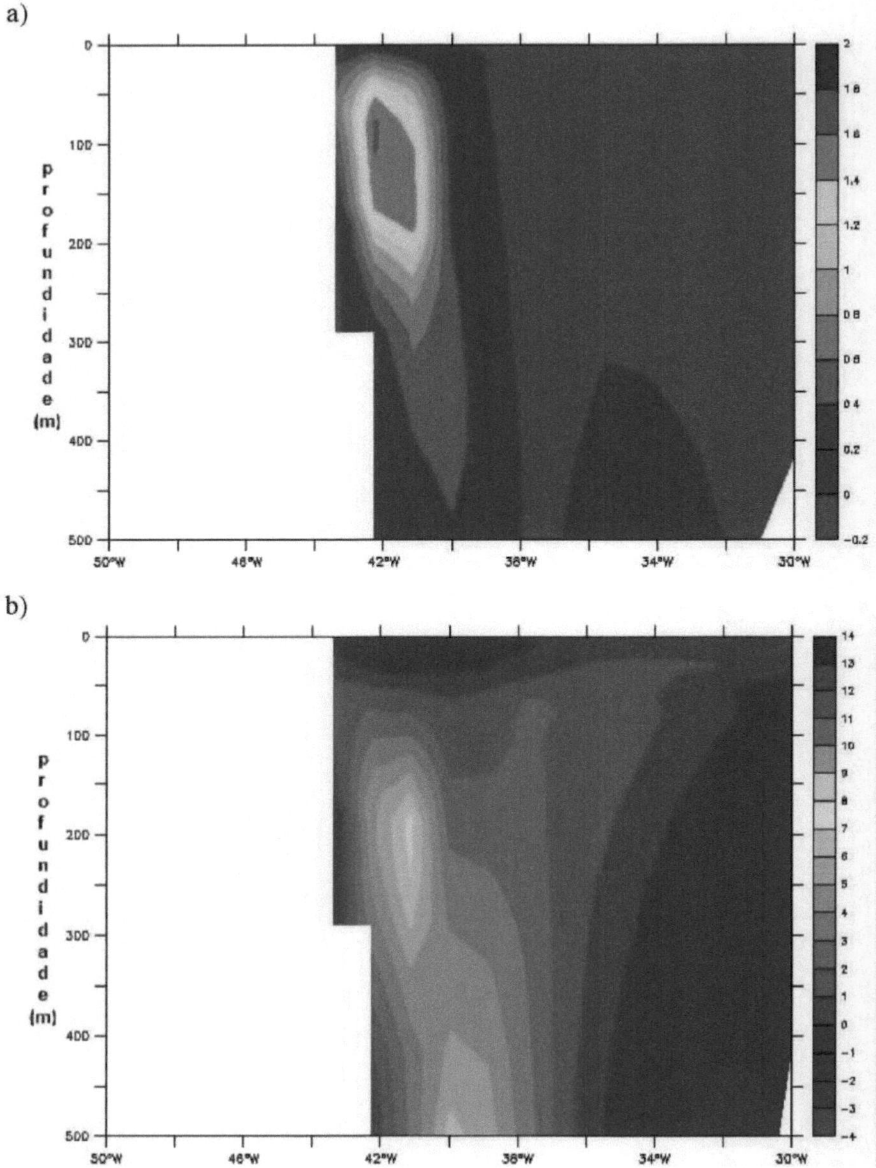

Figure 4.7: Zonal vertical section in the ocean of the vertical velocity of sea water (m.s-1) for: a) the fall of the PI period, b) the anomaly (MH-PI). The scales correspond to values that have been multiplied by $10e^5$ and by 10, respectively.

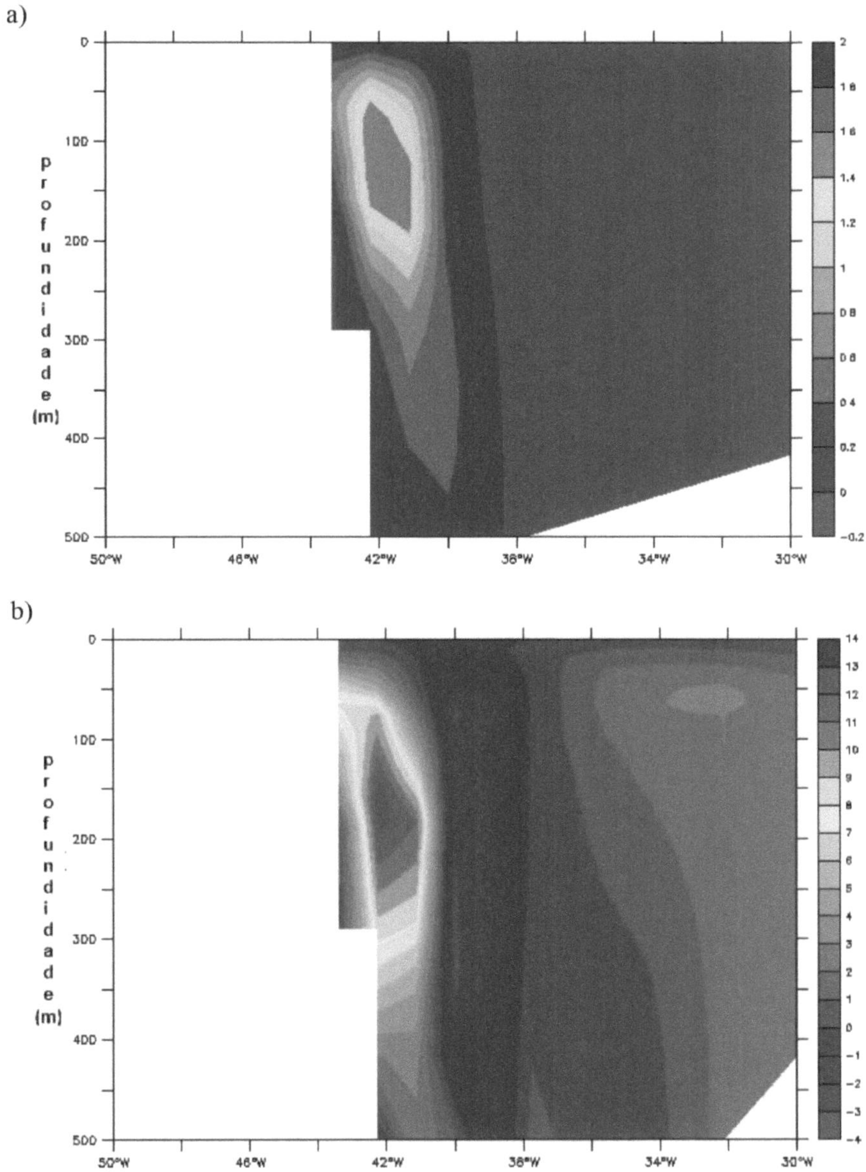

Figure 4.8: Zonal vertical section in the ocean of the vertical velocity of sea water ($m.s^{-1}$) for: a) winter of the PI period, b) anomaly (MH-PI). The scales correspond to values that have been multiplied by $10e^5$ and by 10, respectively.

a)

b)

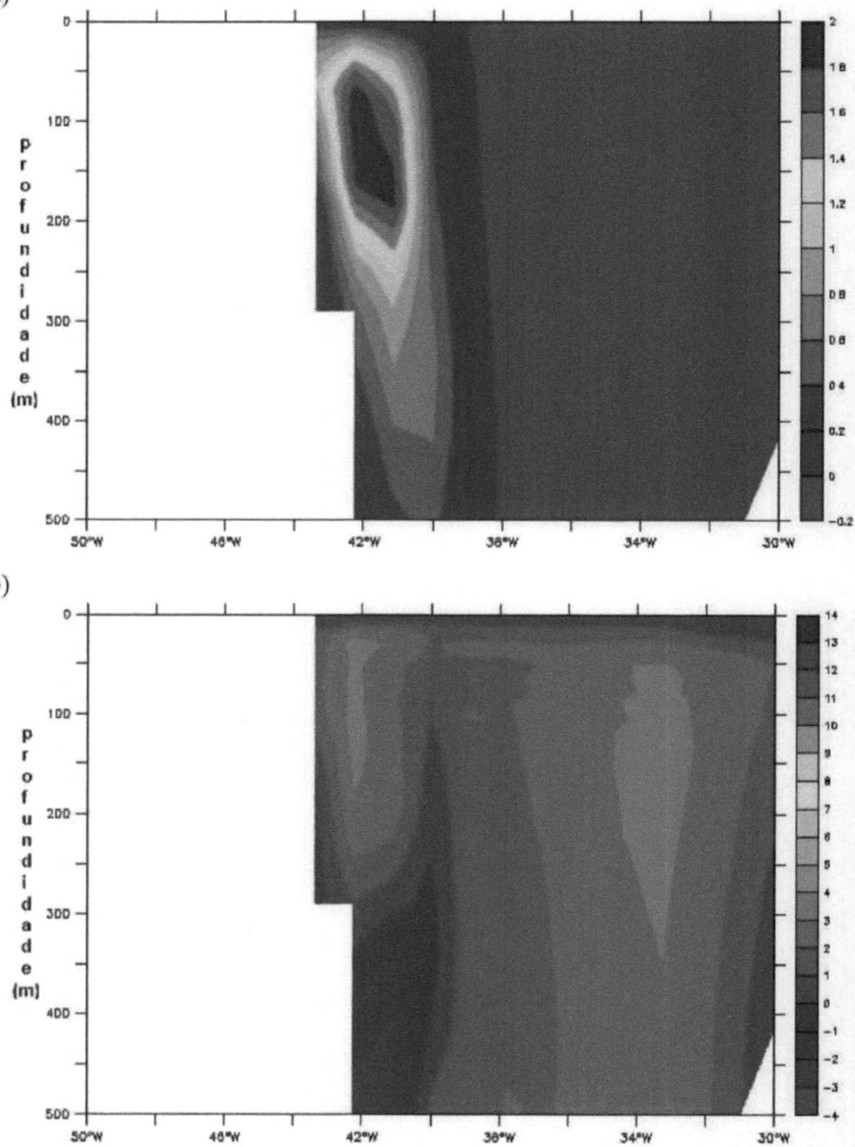

Figure 4.9: Zonal vertical section in the ocean of the vertical velocity of sea water (m.s-1) for: a) spring PI period, b) anomaly (MH-PI). The scales correspond to values that have been multiplied by $10e^5$ and by 10, respectively.

4.2.2. Vertical time distributions

Figures 4.10 a) and b) illustrate the vertical time distributions of the vertical component of the current for the PI and anomaly periods. Seasonal variation can be seen for both the PI and MH periods, with higher values occurring in spring and summer and lower values in late fall to early winter. Through the anomaly, we can see values close to zero at shallower depths, which means that there were small differences between the periods analyzed for these layers. However, this is not the case at greater depths, where there are large positive and negative differences throughout the year. There are more intense anomalous positive values at the end of winter and beginning of spring, which means higher ascending speeds during this period in the MH.

a)

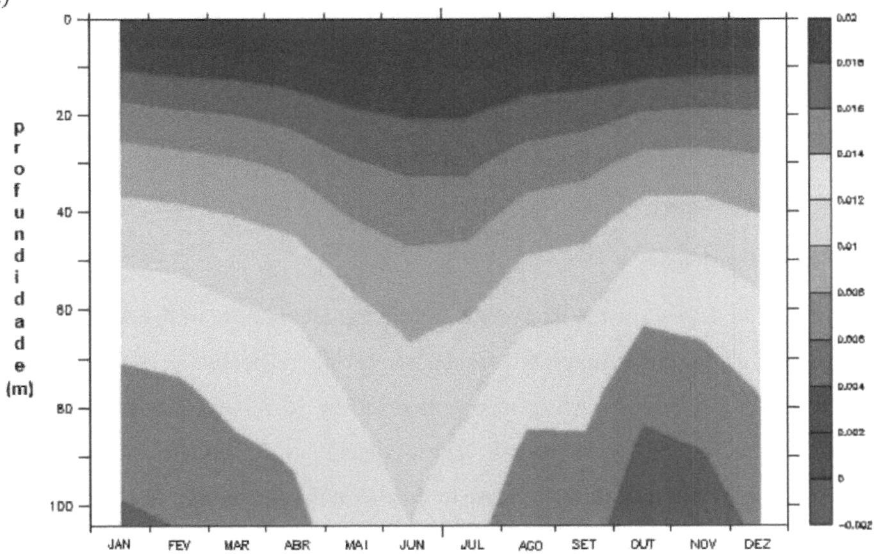

b)

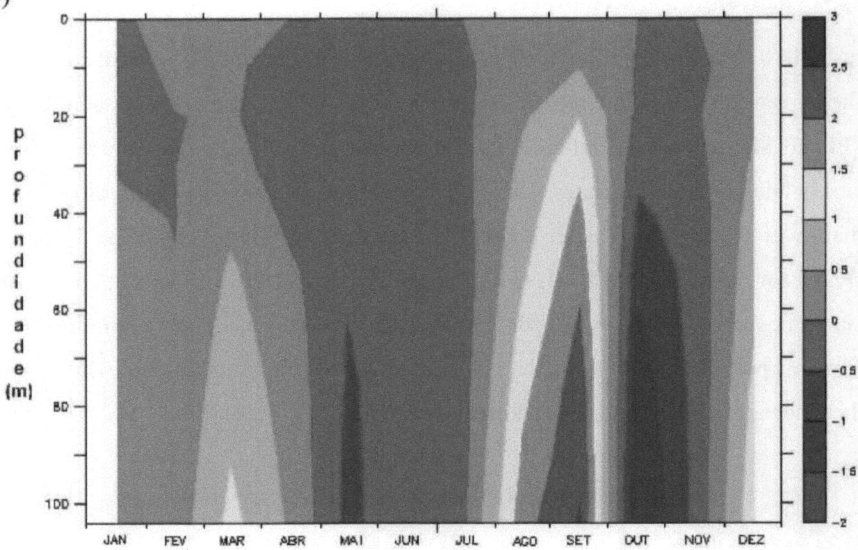

Figure 4.10: Vertical time distribution in the ocean of the vertical velocity of sea water ($m.s^{-1}$) for: a) the PI period, b) anomaly (MH-PI). The scale corresponds to values that have been multiplied by 1000.

4.3. Sea water density

Figures 4.11 to 4.14 show the isopycnals for the top and bottom of the ACAS in zonal vertical sections for the summer, fall, winter and spring, respectively, of the PI and MH periods. The top of the ACAS is represented by the 25.7 isopycnal and the base is represented by the 26.9 isopycnal. These density values were obtained using temperature and salinity values common to the thermohaline indices for the upper limit (180 C and 36) and lower limit (6°C and 34.5) of the ACAS. The black isopycnal refers to the top and the green to the bottom of the ACAS for PI, while the red isopycnal refers to the top and the blue to the bottom of the ACAS for MH. Analysis of these figures reveals that during the MH, the top of the ACAS was closer to the surface, while the base of the ACAS was at greater depths. This indicates a greater influence during the HM of the water mass that resurfaces in Cabo Frio.

Figures 4.15 and 4.16 show the isopycnals for the top of the ACAS in zonal vertical sections for all the seasons of the year in the PI and MH periods respectively. The isopycnals shown in black refer to summer, red to autumn, green to winter and blue to spring. It can be seen that during the PI period the top of the ACAS was closest to the surface during the summer months (at 41 meters), spring (at 42 meters), autumn (at 45 meters) and winter (at 49 meters). During the HM, the top of the ACAS was at a similar depth during spring and summer. However, in the region closest to the coast, the top of the ACAS was shallower during the spring, at a depth of 37.5 meters, while during the summer it was at a depth of 38.5 meters. The top of the ACAS was at its deepest during the fall (at 43 meters) and winter (at 47 meters).

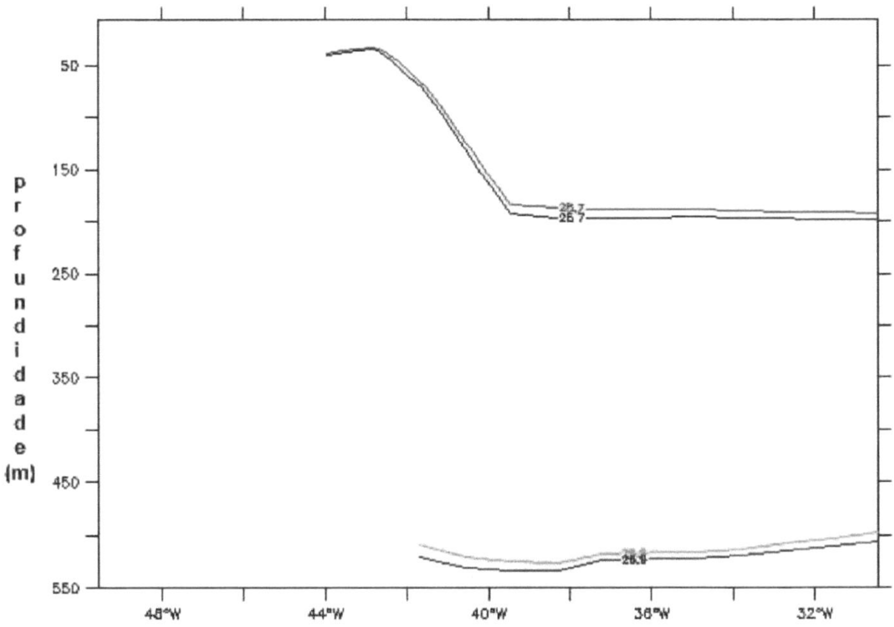

Figure 4.11: Vertical zonal section in the ocean of the density for the summer of the MH and PI periods. The black isopycnal refers to the top and the green to the bottom of the ACAS for PI, while the red isopycnal refers to the top and the blue to the bottom of the ACAS for MH. The top of the ACAS is represented by the 25.7 isopycnal and the base is represented by the 26.9 isopycnal.

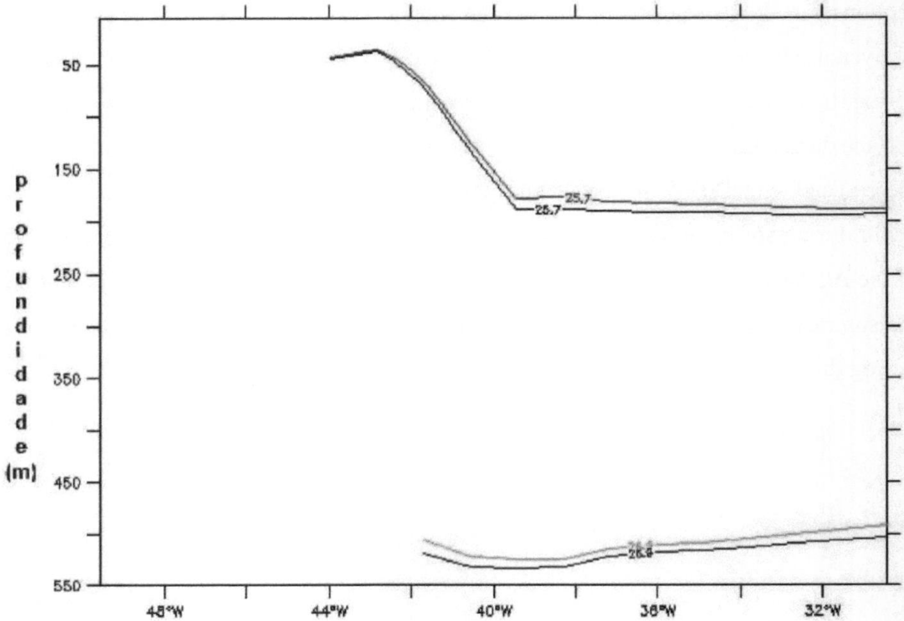

Figure 4.12: Zonal vertical section in the ocean of the density for the fall of the MH and PI periods The black isopycnal refers to the top and the green to the bottom of the ACAS for PI, while the red isopycnal refers to the top of the ACAS and the blue to the bottom of the ACAS for MH. The top of the ACAS is represented by the 25.7 isopycnal and the base is represented by the 26.9 isopycnal.

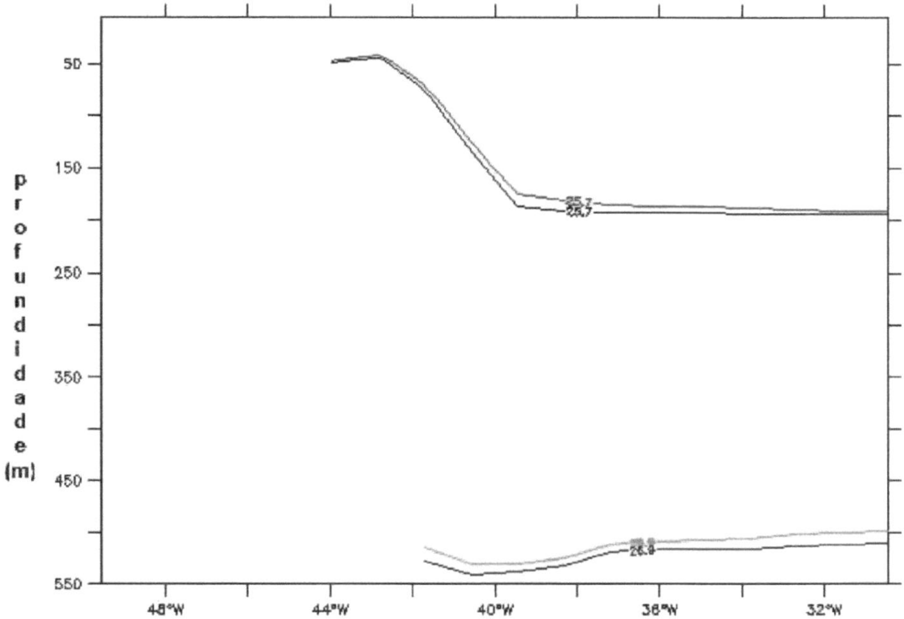

Figure 4.13: Zonal vertical section in the ocean of the density for the winter of the MH and PI periods. The black isopycnal refers to the top and the green to the bottom of the ACAS for PI, while the red isopycnal refers to the top and the blue to the bottom of the ACAS for MH. The top of the ACAS is represented by the 25.7 isopycnal and the base is represented by the 26.9 isopycnal.

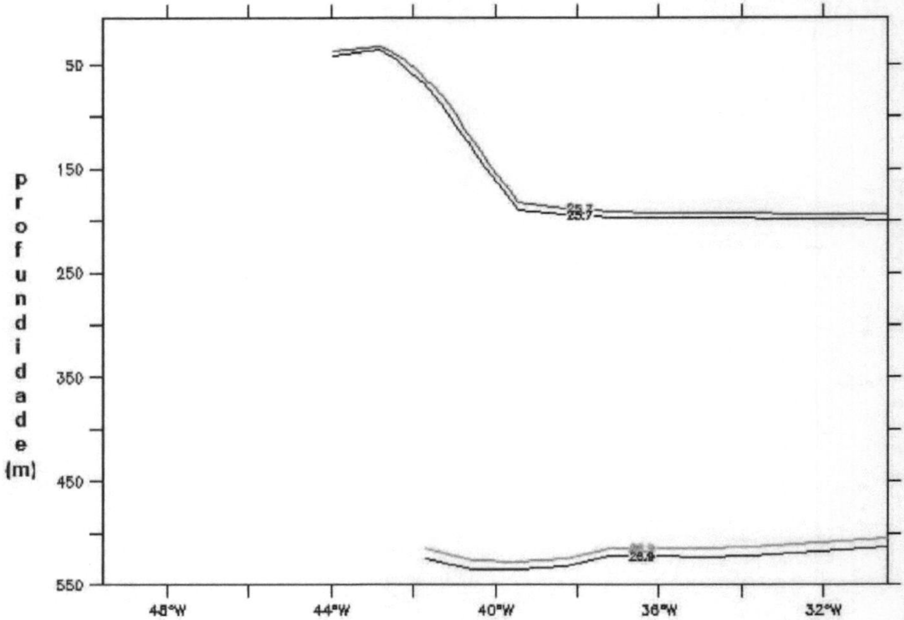

Figure 4.14: Zonal vertical section in the ocean of the density for the spring MH and PI periods. The black isopycnal refers to the top and the green to the bottom of the ACAS for PI, while the red isopycnal refers to the top and the blue to the bottom of the ACAS for MH. The top of the ACAS is represented by the 25.7 isopycnal and the base is represented by the 26.9 isopycnal.

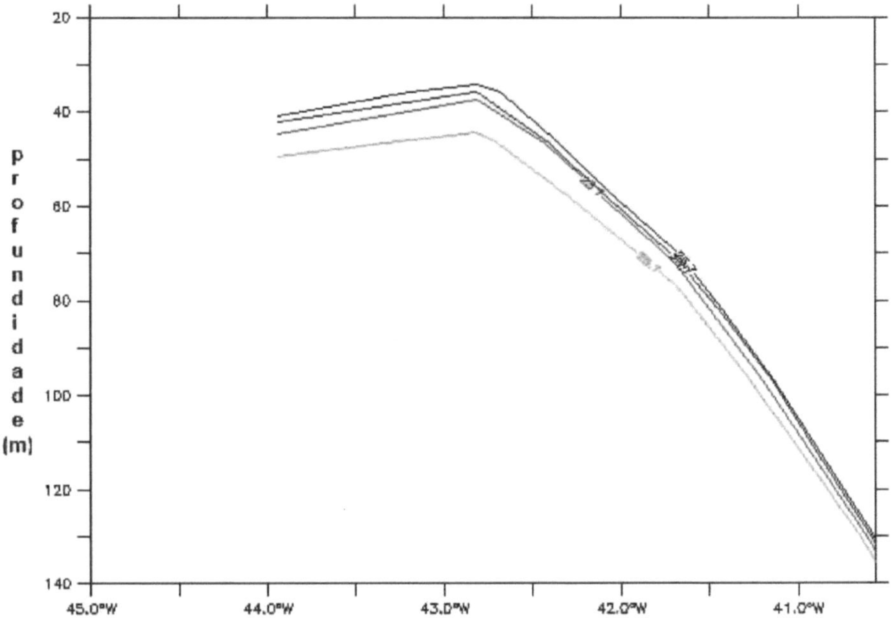

Figure 4.15: Zonal vertical section in the ocean of the density for the PI period. The top of the ACAS is represented by the 25.7 isopycnal which is plotted in black for summer, red for autumn, green for winter and blue for spring.

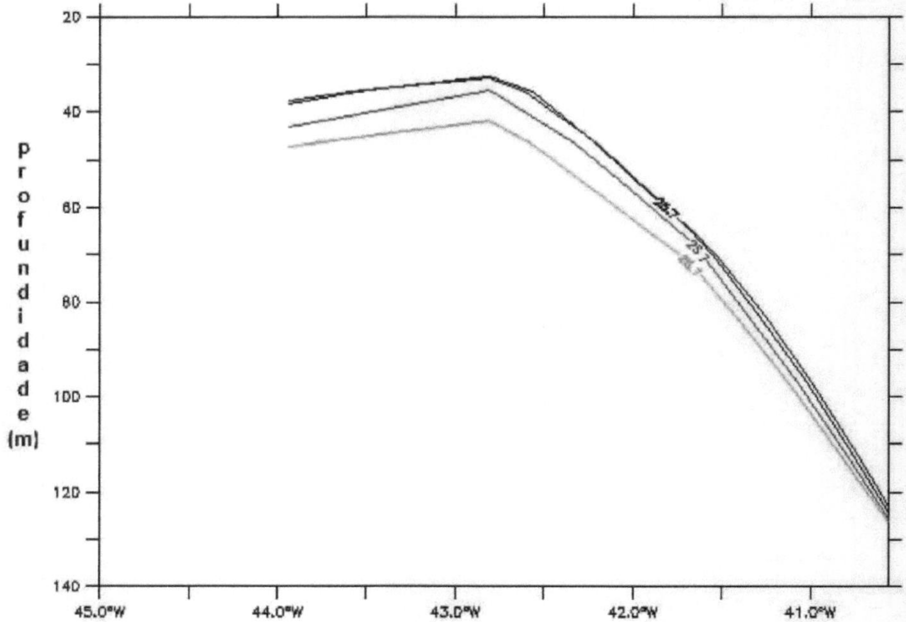

Figure 4.16: Zonal vertical section in the ocean of the density for the MH period. The top of the ACAS is represented by the 25.7 isopycnal which is plotted in black for summer, red for autumn, green for winter and blue for spring.

4.4. Atmospheric pressure at mean sea level

Figures 4.17 to 4.20 show the pressure fields at mean sea level for both the PI and MH periods. The behavior of the SSA and its seasonal variation can be seen. Comparing the experiments, it can be seen that during the summer the SSA was at the same intensity, but closer to the continent during the MH. In the fall, it was at the same intensity, but further away from the continent during the HM. During the HM, it weakens in winter and intensifies in spring.

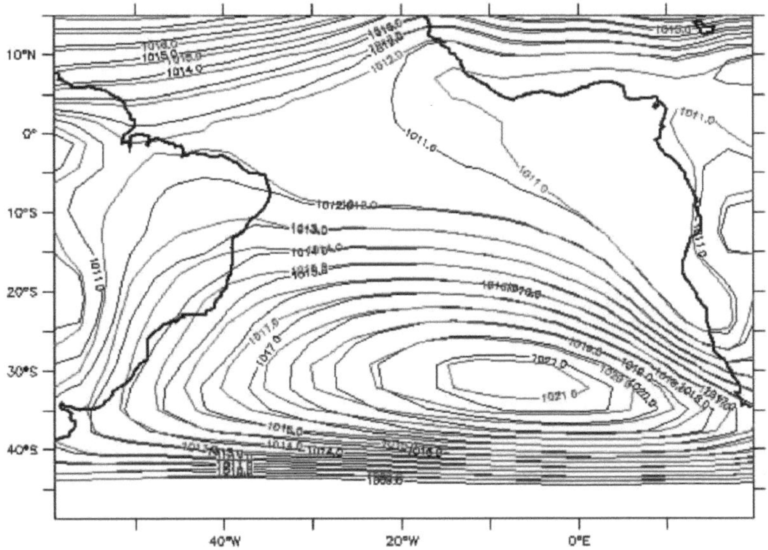

Figure 4.17: Sea level air pressure field (hPa) for the summer PI (black) and MH (red) periods.

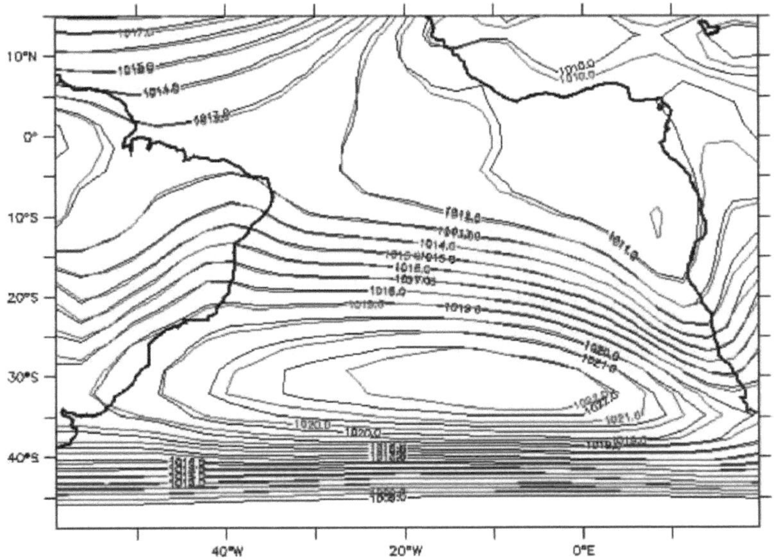

Figure 4.18: Sea level air pressure field (hPa) for the fall PI (black) and MH (red) periods.

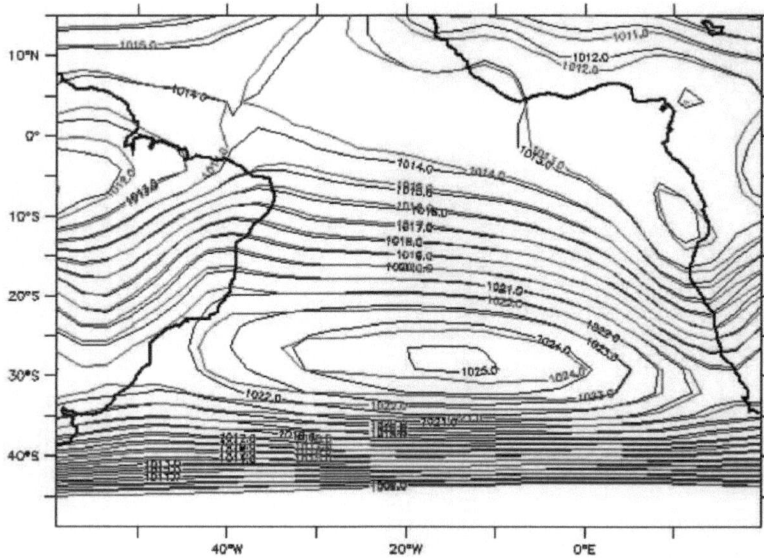

Figure 4.19: Sea level air pressure field (hPa) for the winter PI (black) and MH (red) periods.

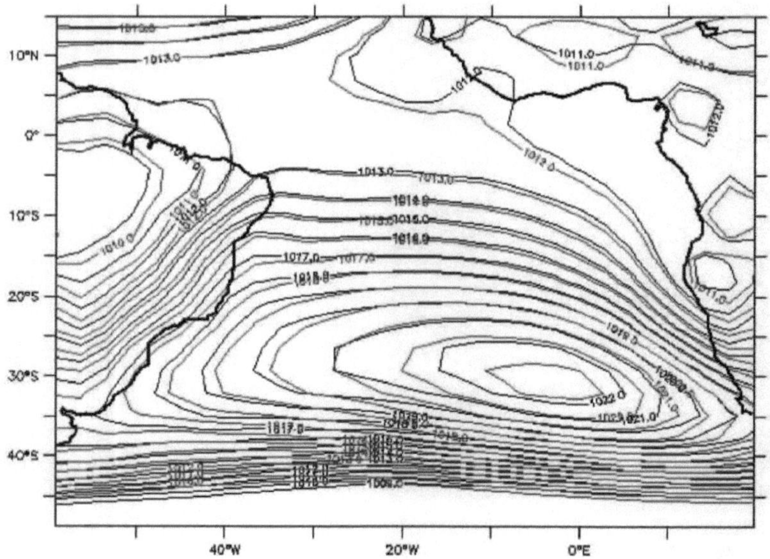

Figure 4.20: Sea level air pressure field (hPa) for the spring PI (black) and MH (red) periods.

4.5. Wind

Figures 4.21 to 4.24 show the wind fields at 925 hPa for the IP period and the anomaly. For MH and PI throughout the year, there is a strong influence of the SSA in the region adjacent to Cabo Frio. Using the wind anomaly fields, we can see that in relation to the circulation of the ASAS, during the MH there was an increase in the mean flow at 925hPa in the summer and a decrease during the winter compared to the PI period. During the fall and spring, there was little variation in the mean flow at 925hPa.

a)

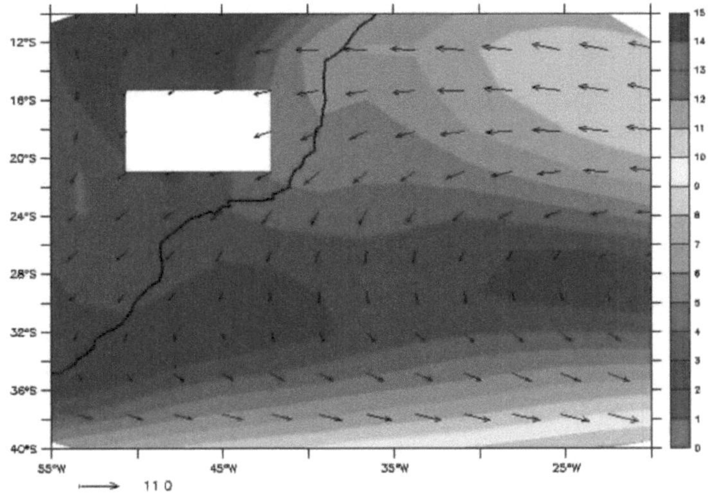

b)

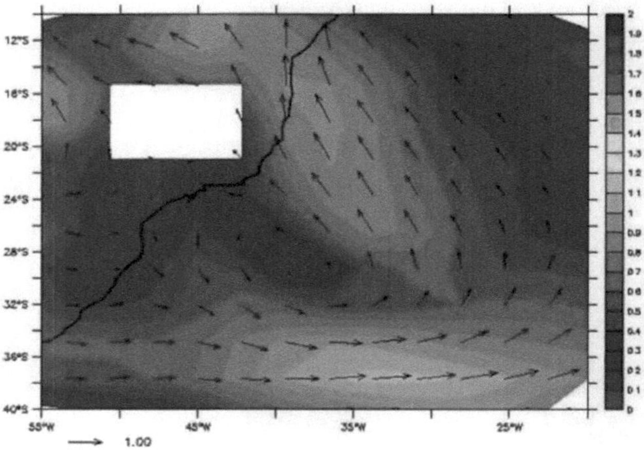

Figure 4.21: Wind field (ms^{-1}) at 925 hPa for: a) PI summer, b) anomaly (MH-PI).

a)

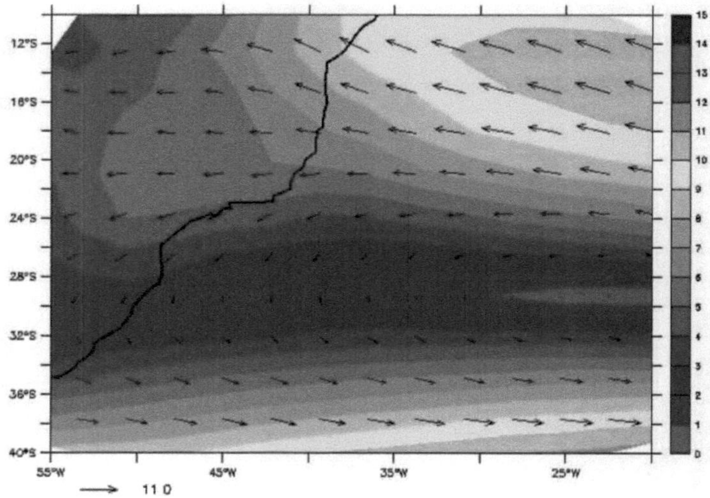

b)

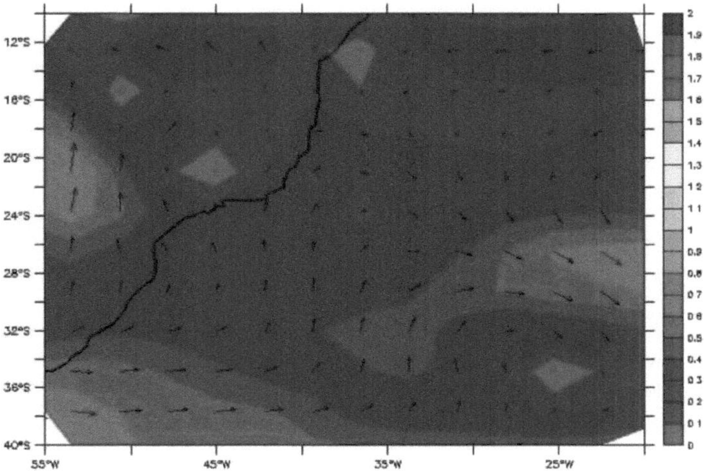

Figure 4.22: Wind field (ms-1) at 925 hPa for: a) fall PI, b) anomaly (MH-PI).

a)

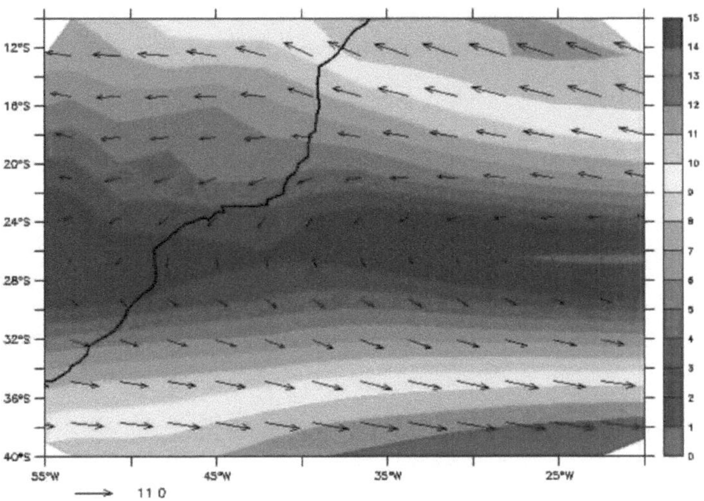

b)

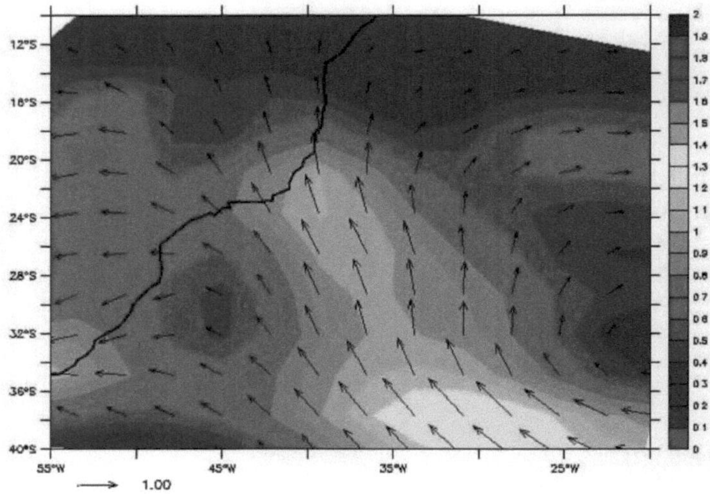

Figure 4.23: Wind field (ms-1) at 925 hPa for: a) PI winter, b) anomaly (MH-PI).

a)

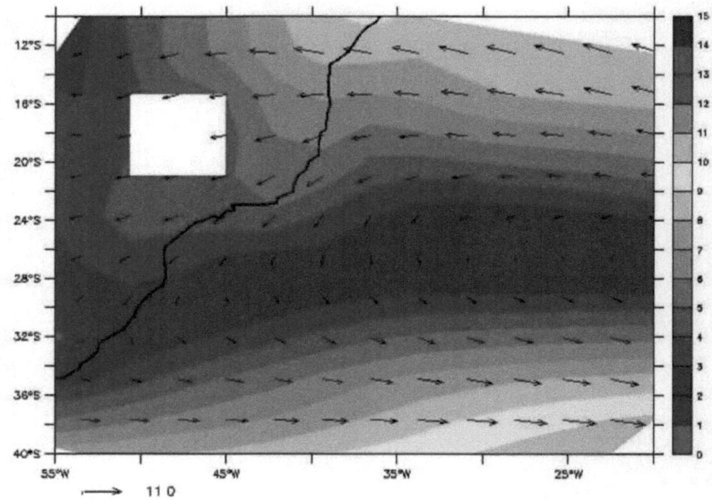

b)

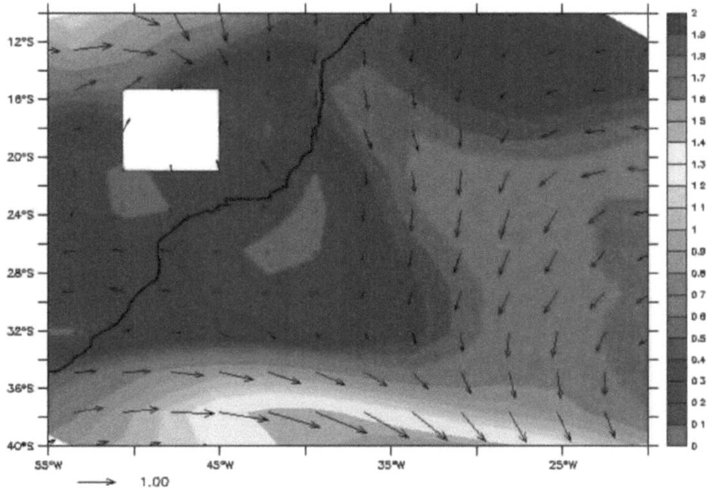

Figure 4.24: Wind field (ms-1) at 925 hPa for: a) PI spring, b) anomaly (MH-PI).

4.6. resurgence index

4.6.1. Surface fields

To analyze the effect of wind shear stress in generating vertical movements in the ocean, we estimated the upwelling index (item 3.3), shown in figures 4.25 to 4.28. According to the formulation of this index, for the Cabo Frio region, the more negative the rotational wind shear stress, the greater the vertical upward velocity at the base of the surface Ekman layer in the ocean, and the more positive the upwelling index. An analysis of these figures shows that the upwelling index is positive in a coastal strip close to the edge of the ASAS during all the seasons of the year in both the PI and MH periods. The most intense values are observed during the spring and summer periods, respectively.

a)

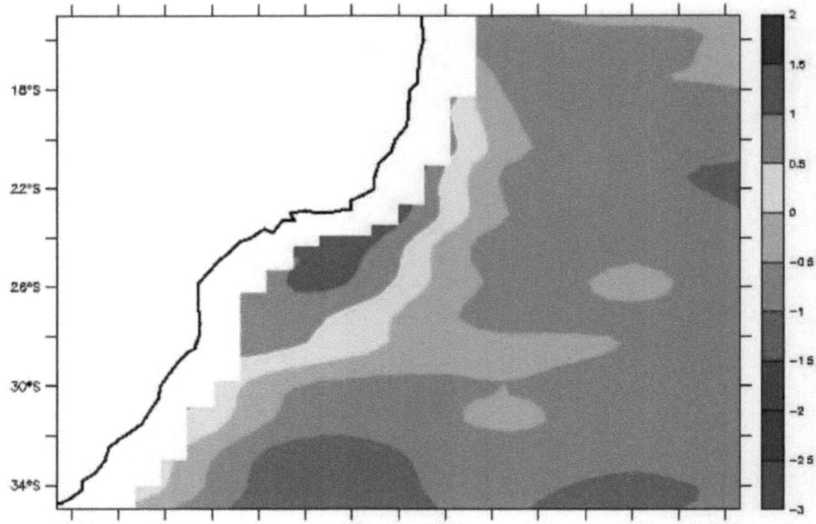

b)

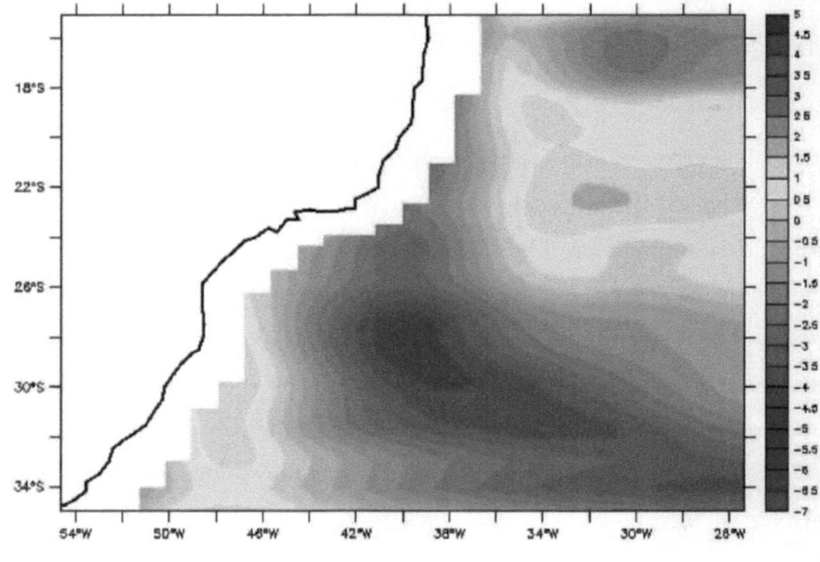

Figura 4.25: Resurgence index field (ms^ ^{1}) for: a) PI summer, b) anomaly (MH-PI). The scale corresponds to values that have been multiplied by 10e^{6} and 10e^{7}, respectively.

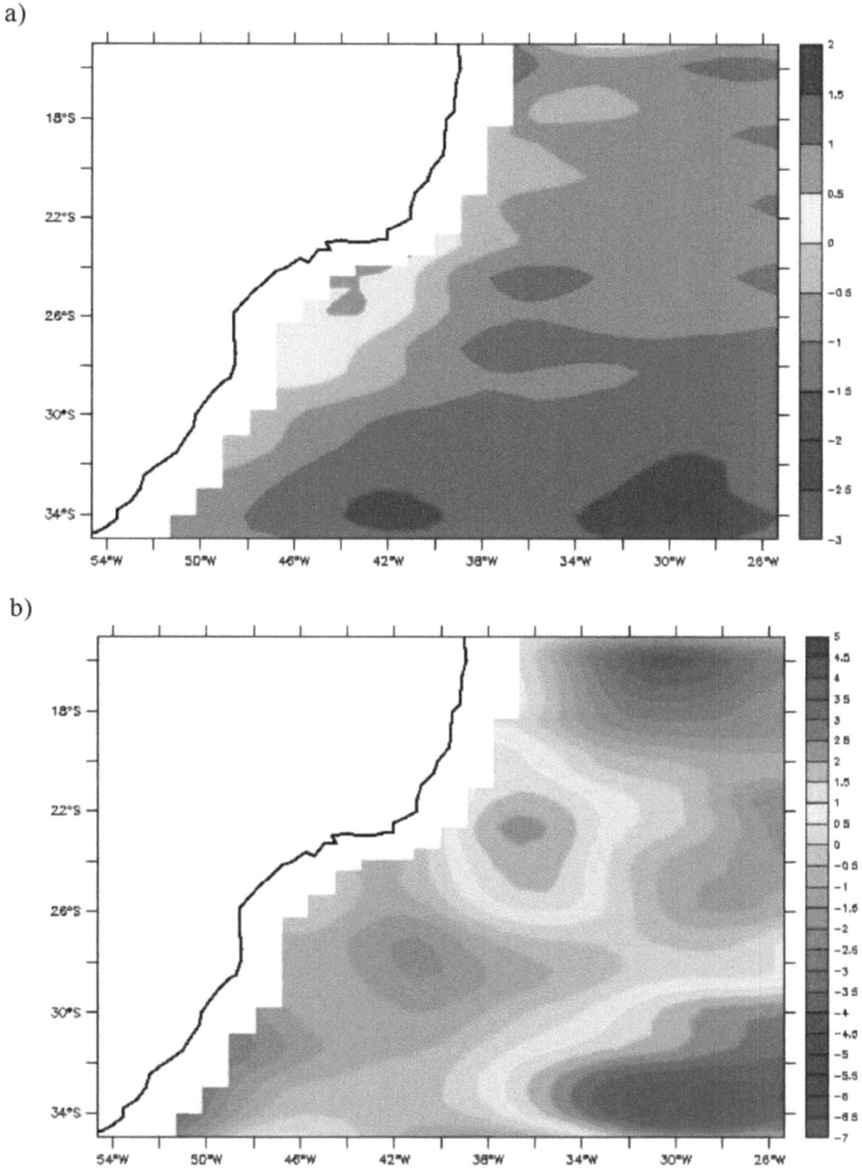

Figura 4.26: Resurgence index field (ms-1) for: a) PI autumn, b) anomaly (MH-PI). The scale corresponds to values that have been multiplied by $10e^6$ and $10e^7$, respectively.

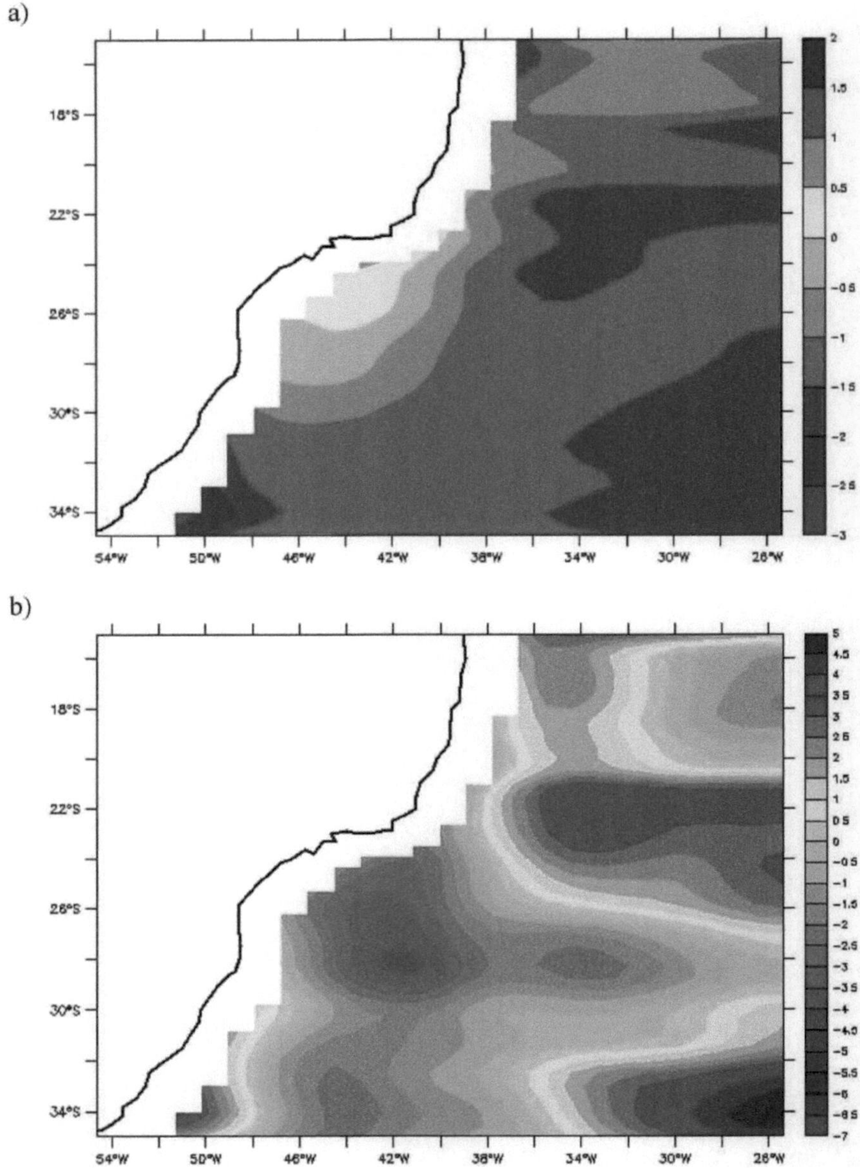

Figura 4.27: Resurgence index field (ms^{-1}) for: a) PI winter, b) anomaly (MH-PI). The scale corresponds to values that have been multiplied by $10e^6$ and $10e^7$, respectively.

a)

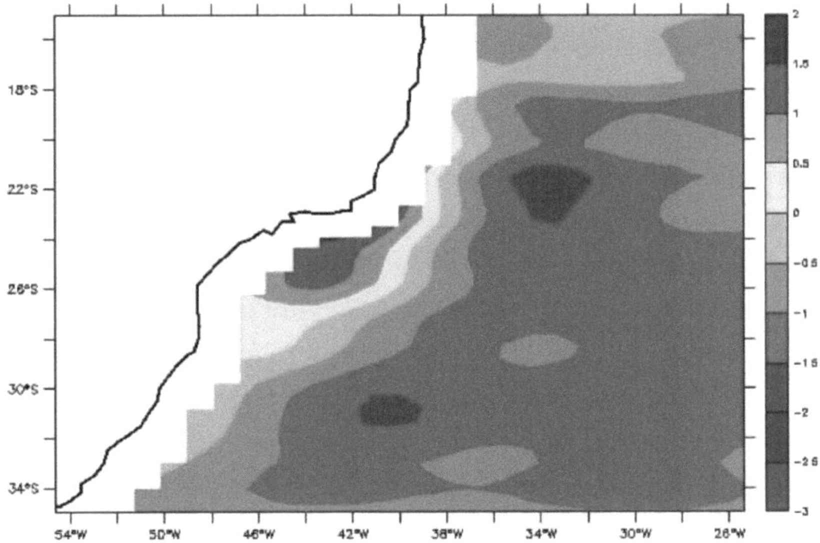

b)

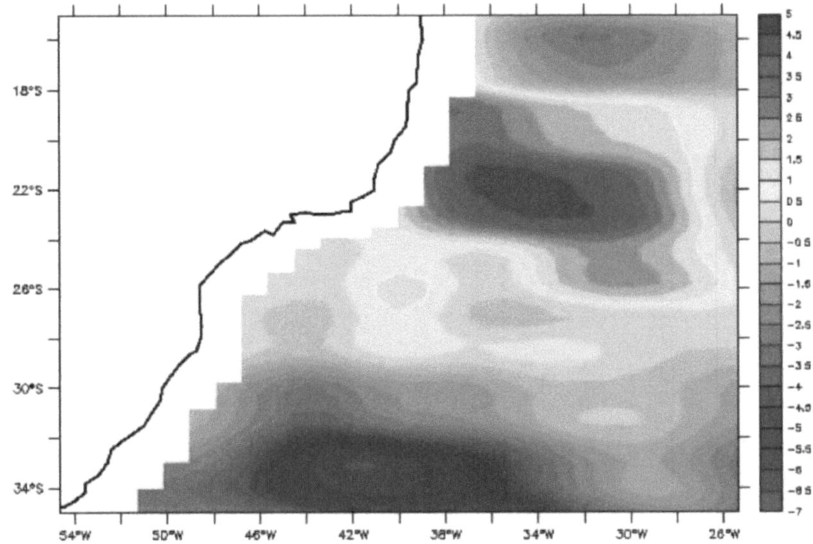

Figura 4.28: Resurgence index field (ms^{-1}) for: a) PI spring, b) anomaly (MH-PI). The scale corresponds to values that have been multiplied by $10e^6$ and $10e^7$, respectively.

4.6.2. Time series

Figure 4.29 shows the time series of the upwelling index for the HM, IP and anomaly. Analysis of this figure shows similar seasonal behavior between the periods, with a correlation of 0.87. The highest values are found during the spring and summer months for both experiments. Positive values can be observed throughout the PI period, while during the MH, negative values occur during the months of May, June and July. These results indicate that for all seasons of the year, the atmospheric circulation in the study region contributes substantially to oceanic processes that favor the pumping of water into the surface layers. Analyzing the time series of the upwelling index anomaly, a positive value can be seen only for the months of September and December, meaning that excluding only these months, higher values of the upwelling index occurred during the PI period.

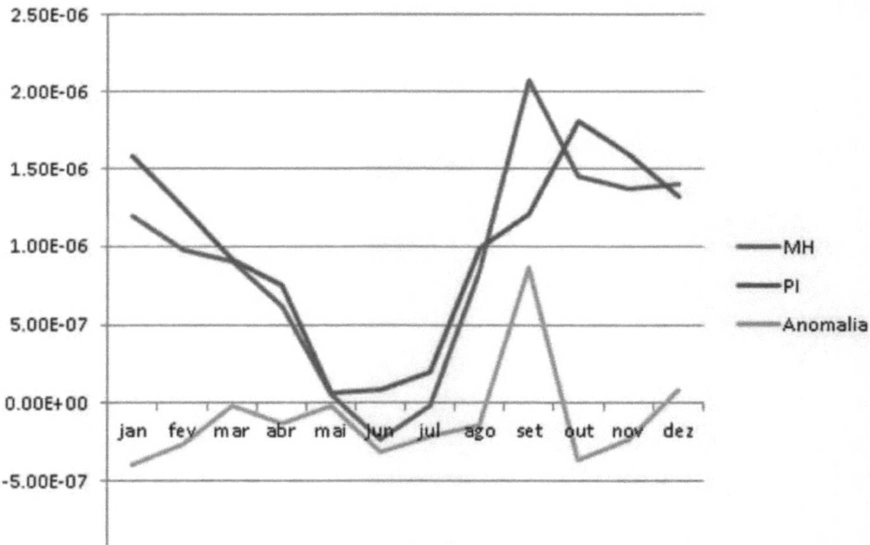

Figure 4.29: Time series of the upwelling index (ms-1) for the MH (blue), PI (red) and anomaly (green) fixed at parallel 23.5S and longitude 41.7W.

There are some limitations to the methods used in the current study. The results of the

CCSM3 simulations have low spatial resolution, making it difficult to analyze coastal ocean processes such as upwelling in detail. Therefore, in order to carry out a more detailed analysis of the phenomenon under investigation, it is necessary to implement a regional ocean model to carry out numerical experiments with high spatial resolution in the Cabo Frio region, allowing the occurrence and intensity of the upwelling process in Cabo Frio to be identified in greater spatial detail. The upwelling in this region is characterized as the rise of the ACAS to the surface, and the intensity of the occurrence of this mass of water on the platform is influenced by the morphology of the coast (Rodrigues and Lorenzzetti, 2001), which was different from the present during the HM due to the higher sea level (Martin *et al.* 1993). Thus, in order to have this effect represented, it would be necessary to use a bathymetry appropriate to the period analyzed.

The experiments carried out following the PMIP 2 protocol use the temperature and salinity from the Levitus 1998 data set as the initial ocean condition for the MH simulation, just as the PI period simulation was initialized, or 100 years after the PI period simulation was carried out.

Large and Danabasoglu (2006) reported some of the biggest and potentially most important errors in CCSM3 in simulating the climate of today, focusing more attention on the SST because it is the most important ocean parameter involved in coupling with the atmosphere. The results obtained suggest that the errors in TSM may be caused by the model underestimating wind stress and overestimating surface radiation. This weaker wind stress results in less intense upwelling restricted to a shallow layer, and thus less cooling of the upper layers of the ocean, which also occurs due to excess surface insolation.

5. CONCLUSIONS

Wind fields at 925 hPa and pressure at mean sea level for all seasons of the year in both experiments showed that the atmospheric circulation was strongly influenced by the ASAS.

Analysis of the vertical seawater density profile revealed that during the HM, the top of the ACAS was closer to the surface, while the base of the ACAS was at greater depths. This indicates a greater influence during the HM of the water mass that resurfaces in Cabo Frio. As observed in the fields of the vertical component of the current, higher values are found in regions closer to the coast. This may be related to the greater divergence of horizontal transport in this region due to the very presence of the coast, and also to the greater occurrence of the upwelling phenomenon in regions close to Cabo Frio (RJ).

Through a seasonal analysis during the HM period, it was identified that during the spring and summer seasons the upwelling index, the vertical velocity of the sea water and the depth of the top of the ACAS were more favorable to the occurrence of upwelling in the Cabo Frio region, while during the fall and winter these parameters were less favorable. These results indicate that during the HM, the upwelling in the Cabo Frio region showed a seasonal behavior, being more intense in spring and summer and weaker in winter and autumn.

Also noteworthy is the possibility of a more intense occurrence of upwelling processes in Cabo Frio during the MH compared to the PI period, during spring and summer. This can be explained by the lower depth of the upper limit of the ACAS and the higher values of the upwelling index and vertical velocities of the sea water during the spring and summer seasons of the MH when compared to the PI period.

In general, the results obtained in this study are in line with the results obtained through paleoclimate indicators. According to studies based on paleoclimatic indicators, during the HM the sea level was higher (Martin *et al.* 1993), which reduced the input of continental organic matter (Mahiques *et al.* 2007; Nagai *et al.* 2009) and would have allowed the main CB flow to move further towards the coast (Mahiques *et al.* 2007).

According to Nagai *et al*. (2009), the occurrence of ACAS on the shelf would result in relatively high

primary productivity and possibly also the occurrence of relatively high velocity bottom currents, which caused low organic carbon content due to resuspension. Episodic events such as El Nino were infrequent during this period (Laslandes *et al*. 2006; Liu, Kutzbach, Wu, 2000; Melo, 2007, Otto-Bliesner *et al*., 2006a).

6. BIBLIOGRAPHICAL REFERENCES

BAKUN, A. Global Climate Change and Intensification of Coastal Ocean Upwelling, in *Science*, p. 198-201, 1990.

BERGER, A.: Long-Term Variations of Caloric Solar Radiation Resulting from the Earth's Orbital Elements, Quatern. Res., 9, 139-167, 1978a

BERGE R, A. L. Notes and Correspondence: Long-Term Variations of Daily Insolation and Quaternary Climatic Changes. Journal of Atmospheric Science. v. 35, p. 2362-2367. 1978b.

BRACONNOT, P.; OTTO-BLIESNER, B.; HARRISON, S.; JOUSSAUME, S.; PETERCHMITT, J.-Y.; ABE-OUCHI, A.; CRUCIFIX, M.; FICHEFET, T.; HEWITT, C.D.; KAGEYAMA, M.; KITOH, A.; LOUTRE, M.-F.; MARTI, O.; MERKEL, U.; RAMSTEIN, G.; VALDES, P.; WEBER, S.L.; YU, Y.; and ZHAO, Y. Results of PMIP2 coupled simulations of the Mid-Holocene and Last Glacial Maximum - Part 1: experiments and large-scale features, Clim. Past, v.3, p. 261-277, 2007.

BRANDINI, F.P. Hydrography and biological production in the Southeast-South region of Brazil in the context of the Revizee program. *In*: Rossi-Wongtschowski, C.L.D.B., Madureira, L.S.-P. (Eds.), O Ambiente oceanogràfico da plataforma continental e do talude na regiao Sudeste-Sul do Brasil. Edusp, 2006. p. 459-466.

BRIGLEB, B.P.; BITS, C.M.; HUNKE, E.C.; LIPSCOMB, W.H.; HOLLAN, M.M.; SCHRAMM, J.L. AND MORITZ, R. E. Scientific description of the sea ice component in the Community Climate System Model, Version 3. NCAR Tech. Note NCAR/ TN-463STR, 2004.

CARBONEL, C.; VALENTIN, J.L. Numerical modelling of phytoplankton bloom in the upwelling ecosystem of Cabo Frio (Brazil). Ecological Modelling , v.116, p. 135148. 1999.

CARBONEL, C.A.A.H.. Modelling of upwelling-downwelling cycles caused by variable wind in a very sensitive coastal system. Continental Shelf Research, v.23, p. 1559-1578. 2003.

COLLINS, W. D.; RASCH, P.J.; BOVILLE, B.A.; HACK, J.J.; MCCAA, J.R.; WILLIAMSON, D.L. and BRIEGLEB , B.P. The formulation and atmospheric simulation of the Community Atmosphere Model version 3 (CAM3). *J. Climate,* v.19, p. 2144-2161. 2006a.

COLLINS, W. D.; BITZ, C.M.; BLACKMON, M.L.; BONAN, G.B.; BRETHERTON, C.S.; CARTON, J.A.; CHANG, P.; DONEY, S.C.; HACK, J.J.; HENDERSON, T.B.; KIEHL, J.T.; LARGE, W.G.; MCKENNA, D.S.; SANTER, B.D. and SMITH, R.D. The Community Climate System Model version 3 (CCSM3). *J. Climate,* v.19, p. 2122-2143. 2006b.

CRUCIFIX, M; BRACONNOT, S.P. AND OTTO-BLIESNER, B. Second phase of Paleoclimate Modelling Intercomparison Project. EOS, V.86, No. 28. 2005.

DICKINSON, R.E.; OLESON, K.W.; BONAN, G.B.; HOFFMAN, F.; THORTON, P.; VERTENSTEIN, M.; YANG, Z.-L. E ZENG, X. The Community Land Model and its climate statistics as a component of the Community Climate System Model. *J. Climate,* v.19, p. 2302-2324. 2006.

DOMINGUEZ, J.M.L.; BITTENCOURT, A.C.S.P.; MARTIN, L. Controls on Quaternary coastal evolution of the east-northeastern coast of Brazil: roles of sealevels history, trade winds and climate. Sed. Geology, v.80, p. 213-232. 1992.

FRANCHITO, S.H.; RAO, V.B.; STECH, J.L.; LORENZZETTI, J.A. The effect of coastal upwelling on the sea-breeze circulation at Cabo Frio, Brazil: a numerical experiment. Annales Geophysicae v.16, p. 866-881, 1998.

GENT, P.R. ; BRYAN, F.O.; DANABASOGLU, G.; LINDSAY, K.; TSUMUNE, D.; HECHT, M.W. E DONEY, S.C. Ocean chlorofluorocarbon and heat uptake during the twentieth century in the CCSM3. *J. Climate,* v.19, p. 2366-2381. 2006.

GILL, A. E. Atmosphere-Ocean Dynamics. 662 pp., 1982.

GRADSTEIN,F.M.; OGG,J.G.; SMITH,A.G.; BLEEKER,W.; LOURENS,L.J. A new Geologic Time Scale, with special reference to Precambrian and Neogene. Episodes, v. 27(2), p. 83-100, 2004.

HALL, N.M.J. E VALDES, P.J. A GCM simulation of the climate 6000 years ago. J. Climate, v.10, p. 3-17, 1997.

IMBRIE, J.; BOYLE, E.A.; CLEMENS, S.C.; DUFFY, A.; HOWARD, W.R.; KUKLA, G.; KUTZBACH, J.; MARTINSON, D.G.; MCINTYRE A.; MIX, A.C.; MOLFINO, B.; MORLEY, J.J.; PETERSON, L.C.PISIAS, N.G.; PRELL, W.L.; RAYTOO, M.E.; SHACKLETONS, N.J. AND TOGGWEILER, J.R. On the structure and origin of major glaciation cycles. 1. Linear responses to Milankovich forcing. *Paleoceanography*, v.7, p. 701-738, 1992.

LARGE, W. G., AND DANABASOGLU, G. Attribution and impacts of upper-ocean biases in CCSM3. *J. Climate,* v.19, p. 2325-2346. 2006.

LASLANDES, B.; SYLVESTRE, F.; SIFEDDINE, A.; TURCQ, B.; ALBUQUERQUE, A.L.S.; ABRAÔ, J. Enregistrement de la variabilité hydroclimatique au cours des 6500 dernières années sur lê littoral de Cabo Frio (Rio de Janeiro, Brésil). Comptes Rendus Geoscience, v.338, p. 667-675, 2006.

LEDRU, M.; SALGADO-LABOURIAU, M.L.; LORSCHEITTER, M.L. Vegetation dynamics in southern and central Brazil during the last 10,000 a BP. Review of Palaeobotany and Palynology, v.99, p. 131-142, 1998.

LIU, Z.Y.; KUTZBACH, J. AND WU, L.X. Modeling climate shift of El Nino variability in the Holocene, Geophys. Res. Lett., v.27(15), p. 2265 - 2268, 2000.

LIU , Z; HARRISON , S.P.; KUT ZBAC H, J.; OTTO -BLIESNE R, B. Global monsoons in the mid-Holocene and oceanic feedback. Climate Dynamics. p. 157182, 2004.

MAHIQUES, M.M.; BiCEGO, M.C.; SILVEIRA, I.C.A.; SOUSA, S.H.M.; LOURENÇO, R.A.; FUKUMOTO, M.M. Modern sedimentation in the Cabo Frio upwelling system, Southeastern. Brazilian shelf. Anais da Acadêmia Brasileira de Ciências, v.77 (3), p. 535-548, 2005.

MAHIQUES, M.M.; FUKUMOTO, M.M.; SILVEIRA, I.C.A.; FIGUEIRA, R.C.L.; BiCEGO, M.C.; LOURENÇO, R.A.; SOUSA, S.H.M. Sedimentary changes on the

Southeastern Brazilian upper slope during the last 35,000 years. Anais da Academia Brasileira de Ciências, v.79 (1), p. 171-181, 2007.

MARTIN, L.; FLEXOR, J.M. E VALENTIN, J.L. Influence of the peaceful oceanic phenomenon, El Nino', on the upwelling and climate of the Cabo Frio region, on the Brazilian coast of the State of Rio de Janeiro, *C. R. Acad., Sci. Paris,* v.*307, Série II*, p. 1101-1105, 1988.

MARTIN, L.; ABSY, M.L.; FLEXOR, J.M.; FOURNIER, M.; MOURGUIART, P.; SIFEDDINE, A. AND TURCQ, B. Southern oscillation signal in South American palaeoclimatic data of the last 7000 years. Quaternary Research, v.39, p. 338-46, 1993

MARTINS, L.R. E COUTINHO, P.N. The Brazilian continental margin. *Earth Science Reviews*, v.17, p. 87-107, 1981.

MELO, M. L. D. Climate simulations for the Middle Holocene using CPTEC's MCGA, with emphasis on South America, Instituto Nacional de Pesquisas Espaciais, Tese (Doutorado), 2007.

MELO, M. L. D. and MARENGO, J. A. Simulations of the Middle Holocene climate in South America with the CPTEC general circulation model. Revista Brasileira de Meteorologia, v.23, n.2, 191-205. 2008.

MIRANDA, L.B. Form of the T-S correlation of coastal and oceanic water masses between Cape St. Thomas (RJ) and Sao Sebastiao Island (SP), Brazil. Boletim do Instituto Oceanogràfico, v.33 (2), p. 105-119, 1985.

NAGAI, R.H.; SOUSA, S.H.M.; BURONE, L.; MAHIQUES, M.M. Paleoproductivity changes during the Holocene in the inner shelf of Cabo Frio, southeastern Brazilian continental margin: Benthic foraminifera and sedimentological proxies. Quaternary International, v. 206, p.62-71 , 2009.

JOSSAUME, S. and TAYLOR, K. E. Status of the paleoclimate Modeling Intercomparison Project. In Proceedings of the First International AMIP Scientifc Conference, WCRP (World Climate Researche Programme) Report, p. 425 - 430, 1995.

OTTO-BLIESNER, B.; BRADY, E.; CLAUZET, G.; TOMAS, R.; LEVIS, S.; KOTHAVALA, Z. Last Glacial Maximum and Holocene climate in CCSM3. Journal of Climate, v.19, n. 11, p. 2526-2544. 2006a.

RAMSTEIN, G.; KAGEYAMA, M.; GUIOT, J.; WU, H.; HÉLY, C.; KRINNER, G.; BREWER, S. How cold was Europe at the last glacial maximum? A synthesis of the progress achieved since the first PMIP model-data comparison, Clim. Past 3 (2007), p. 331-339. 2007.

RAYNAUD, D.; JOUZEL, J.; BARNOLA, J. M.; CHAPPELLAZ, J.; DELMAS, R.J., and LORIUS, C. The ice record of greenhouse gases. *Science,* v.259, p. 926934, 1993.

RIEBESELL, U. Effects of CO_2 enrichment on marine phytoplankton. Journal of Oceanography, v.60, p. 719-729, 2004.

RODRIGUES, R.R.; LORENZZETTI, J.A. A numerical study of the effects of bottom topography and coastline geometry on the Southeast Brazilian coastal upwelling. Continental Shelf Research, v.21, p. 371-394, 2001.

SCHEEL-YBERT, R. Vegetation stability in the southeastern Brazilian coastal area from 5500 to 1400 ^{14}C deduced from charcoal analysis. Review of Palaeobotany and Palynology, v.110, p. 111-138, 2000.

SILVEIRA, I.C.A.; SCHIMIDT, A.C.K.;CAMPOS, E.J.D.; GODOI, S.S. & IKEDA, Y.2000. The Brazil Current off the Brazilian East Coast. Revista Brasileira de Oceanografia, **48** (2): 171-183

SVERDRUP, H.U.; JOHNSON, M. W. and FLEMING, R.H. The Oceans. New York. Prentice Hall, Inc., 1942.

TANAKA, K. Simulação da Ressurgência Comparada com Dados Oceanograficos e de Sensores Remotos em Cabo Frio Local, Instituto Nacional de Pesquisas Espaciais, Tese (Mestrado), 1977.

TASCHETTO, A.S. and WAINER, I. Climatological study of sea surface parameters in the Southwest Atlantic. Revista Brasileira de Meteorologia, v.17, n.2, 229-242, 2002.

TORRES JR., A.R. Numerical study of atmospheric teleconnection between oceanic phenomena in the equatorial Pacific and South Atlantic. Rio de Janeiro: Federal University of Rio de Janeiro, Thesis (Doctorate), 2005.

VALDES, P.J. South American paleoclimate model simulations: how reliable are the models? Journal of Quaternary Science, v.15, n. 4, p. 357-368, 2000.

VALENTIN, J.L.; ANDRÉ, D.L.; JACOB, S.A. Hydrobiology in the Cabo Frio (Brazil) upwelling: two-dimensional structure and variability during a wind cycle. Continental Shelf Research, v.7 (1), p. 77-88, 1987.

VETTORETTI, G.; PELTIER, W.R.; MCFARLANE, N.A. Simulations of midHolocene climate using an atmospheric general circulation model. Journal of Climate, v.11, p. 2607-2627. 1998.

MIX
Papier aus verantwortungsvollen Quellen
Paper from responsible sources
FSC® C105338

Printed by Books on Demand GmbH, Norderstedt / Germany